Astronomers' Universe

For other titles published in this series, go to
http://www.springer.com/series/6960

Sun Kwok

Stardust

The Cosmic Seeds of Life

 Springer

Property of Library
Cape Fear Community College
Wilmington, NC

Sun Kwok
The University of Hong Kong
Faculty of Science
Hong Kong
China, People's Republic

ISSN 1614-659X
ISBN 978-3-642-32801-5   ISBN 978-3-642-32802-2 (eBook)
DOI 10.1007/978-3-642-32802-2
Springer Heidelberg New York Dordrecht London

Library of Congress Control Number: 2013933988

© Springer-Verlag Berlin Heidelberg 2013
This work is subject to copyright. All rights are reserved by the Publisher, whether the whole or part
of the material is concerned, specifically the rights of translation, reprinting, reuse of illustrations,
recitation, broadcasting, reproduction on microfilms or in any other physical way, and transmission or
information storage and retrieval, electronic adaptation, computer software, or by similar or dissimilar
methodology now known or hereafter developed. Exempted from this legal reservation are brief excerpts
in connection with reviews or scholarly analysis or material supplied specifically for the purpose of being
entered and executed on a computer system, for exclusive use by the purchaser of the work. Duplication
of this publication or parts thereof is permitted only under the provisions of the Copyright Law of the
Publisher's location, in its current version, and permission for use must always be obtained from
Springer. Permissions for use may be obtained through RightsLink at the Copyright Clearance Center.
Violations are liable to prosecution under the respective Copyright Law.
The use of general descriptive names, registered names, trademarks, service marks, etc. in this
publication does not imply, even in the absence of a specific statement, that such names are exempt
from the relevant protective laws and regulations and therefore free for general use.
While the advice and information in this book are believed to be true and accurate at the date of
publication, neither the authors nor the editors nor the publisher can accept any legal responsibility for
any errors or omissions that may be made. The publisher makes no warranty, express or implied, with
respect to the material contained herein.

Printed on acid-free paper

Springer is part of Springer Science+Business Media (www.springer.com)

Preface

When I was in my second year of undergraduate studies, I read a book by Fred Hoyle called "*Frontiers of Astronomy*". Before reading this book, I had an idea that astronomy involved observing the skies and monitoring celestial events. From Hoyle's book, I realized that modern astronomy is much more than just observations. It is about applying our knowledge of physics to understand the Universe. As a result, I changed my major from engineering to physics with the goal of becoming an astronomer. As a graduate student at the University of Minnesota, I had the good fortune to witness the beginning of infrared spectroscopy. The mid-infrared detectors developed at Minnesota allowed the exploration of the sky in the infrared, and the unexpected discovery of infrared emissions from old stars led to the first positive identification of a mineral in stardust - silicates.

This book is about 40 years of history of the search for an understanding of the nature of stardust. No one predicted the existence of organic stardust, and certainly no one foresaw the wide spread of organic matter in the Universe. This is a fascinating story that ought to be told.

In the early 1990s I began writing a series of popular articles for the *Sky and Telescope*, *Astronomy*, and *Mercury* magazines. These writings got me interested in writing about science for the general public. This led to my first popular science book *Cosmic Butterflies* published by Cambridge University Press in 2001. The subsequent book tours and invitations to speak in USA and Canada allowed me to meet face to face with many of the readers. The strong interest and thirst for information by the public have convinced me of the need for the communication of the latest scientific results in an authoritative but understandable manner.

When we read about significant discoveries of the past, we often don't appreciate how difficult the path has been. Accounts are often sanitized and simplified. But the reality of science is that success occurs after many errors, detours, and dead ends and is never straightforward. My own participation in the research on this subject has also allowed me to witness first-hand how things happened and I hope to relay these events in this book.

Since science is a human endeavor, personalities are an integral part of the process. In this book, I benefited from the personal accounts of many people involved in the research on stardust, in particular those who related to me their personal experiences on the road to discovery.

Unlike most popular science books, this book is more than a report of discoveries. Through the reading of the primary literature and personal interactions with the scientists who do the work, I was able to evaluate the evidence, form my own critical assessment of the work, and determine how it fit into the overall picture of the development of the field.

On the personal side, I am grateful to NASA and ESA who allowed me access to their telescopes through the policy of open competition for telescope time. Without this generous policy, it would not have been possible for me to contribute to this field.

Both astrochemistry and bioastronomy are new scientific disciplines. My service as a member of the executive committees in the International Astronomical Union astrochemistry and bioastronomy commissions gave me the opportunity to meet other scientists in the field and to promote these two subjects in the general scientific community. This book gives me a way to "wave the flag" and hopefully encourage more young people to pursue research in these areas.

I started the earliest draft of this book more than 10 years ago. Due to my administrative duties, I have only been able to write in bits and pieces of spare time that I can find. I want to thank the people and organizations in different parts of the world who have invited me to give talks on the subject of stardust, which gave me confidence that this subject is indeed of wide public interest.

I want to thank Agnes Lam who kindly gave permission to me to include her beautiful poem in this book, as well as giving valuable comments on an earlier draft of the book. I would like to thank my editor Ramon Khanna of Springer who took an interest in this project. I would also like to express my gratitude to Arturo Manchado for his hospitality during my stay at the Instituto de Astrofísica de Canarias. Anisia Tang, my friend and colleague, helped in the production of some of the drawings used in this book. I want to thank my wife, Emily, and my daughter, Roberta who have read various drafts of this book and gave me valuable feedback and comments.

Hong Kong,   Sun Kwok
People's Republic of China
December 2012

Contents

List of Figures

List of Tables

List of Boxes

Prologue

The Heaven and Earth connection is one of the oldest concepts of mankind. All the ancient cultures subscribed to the belief that our lives are guided and governed by celestial objects. Astrology is just one example of such beliefs. However, with the growth of technology, our connections to the heavens have diminished. With artificial lighting, we are less dependent on the rise and setting of the Sun. The role of the Moon as an illumination source at night is all but forgotten. An increasing number of people live in cities where light pollution makes it difficult for residents to see and appreciate the stars. The passing of comets is something we read in the news, but not the first-hand visual spectacle that awed the citizens of the past.

In modern times, the intellectual community has come to believe that we originated and developed from this Earth. Life began, evolved, and prospered on this planet, in total isolation from the rest of the Universe. Stars are remote, distant, and irrelevant entities. It is in this context and background that I am writing this book, to remind us that stars have been a major part of our origin. We can be oblivious to their birth, life, and death, but it is quite likely that these distant objects were responsible for our existence. If someone were to say this 30 years ago, the idea would have been dismissed out of hand. But lots of things have changed. The development of space and astronomical technology has brought us unprecedented capabilities to study stars. The discovery of stardust, in particular that made up of organics, was totally unexpected and still difficult to understand. In spite of our lack of theoretical understanding, the observational facts are clear and definite. Stars, near the end of their lives, are able to synthesize extremely complex organic compounds under near vacuum conditions. Large quantities of organics are manufactured over very short time scales, and ejected and distributed throughout the Galaxy. With space spectroscopic observations, we can determine the chemical composition of these stardust particles, and surprisingly, we found them to show remarkable resemblance to the organic solids in meteorites. Since meteorites are remnants of primordial solar nebula, is it possible that stars have enriched our Solar System with organics? This idea has gained support from the discovery of pre-solar grains, inorganic stellar grains that have been demonstrated to have come from old stars outside of the Solar System. Recent research has also told us that the Earth was subjected to heavy bombardments from comets and asteroids during the early history of the Earth. These bombardments may have brought with them the primordial organics, seeding the Earth with raw materials as basic ingredients of life.

This scenario was developed as the result of the work of many people. There are astronomers who perform observations of distant stars, laboratory chemists who identify the spectral signatures of organics, space scientists who send probes to comets, asteroids, and planetary satellites, meteoritic scientists who examine the chemical composition of meteorites and interplanetary dust particles, geologists who study the early history of the Earth, and biologists who weave a picture of how life could originate from these distant organics. It has been a very exciting experience for me to have been a part of these teams. Sometimes these discoveries seem too fantastic to be true and there has not been a lack of skeptics in the scientific community.

The question of the origin of life is such a complicated issue that the complete answer may not be secured in the near future. But what we have learned is that we have to keep an open mind for unexpected discoveries and entertain new possibilities resulting from these new findings. What I am certain of is that the final answer will not be arrived at by a scientist from a single discipline, but by teams of scientists attacking the problem from a variety of angles, each bringing a piece of the puzzle that hopefully can be put together to form a picture.

This book is about stardust, the smoke from stellar chimneys. We tell the story of how it was discovered, what it is made of, and what effects it may have on the Solar System and the origin of life.

About the Author

Sun Kwok is a leading world authority on the subject of astrochemistry and stellar evolution. He is best known for his theory on the origin of planetary nebulae and the death of Sun-like stars. His recent research has been on the topic of the synthesis of complex organic compounds in the late stages of stellar evolution. He is the author of many books, including *The Origin and Evolution of Planetary Nebulae* (2000), *Cosmic Butterflies* (2001), *Physics and Chemistry of the Interstellar Medium* (2007), and *Organic Matter in the Universe* (2012). He has been a guest observer on many space missions, including the *Hubble Space Telescope* and the *Infrared Space Observatory*. He currently serves as the President of Commission 34 interstellar Matter of the International Astronomical Union (IAU), as well as Vice President of IAU Commission 51 Bioastronomy. He served as the chairman of IAU Planetary Nebulae Working Group between 1994 and 2001, and as organizing committee member of IAU Astrochemistry Working Group.

Vanilla in the Stars

By Agnes Lam
Special Mention Award, 24th Nosside International Poetry Prize

When I was a child,
I used to gaze at the stars above

our garden of roses, jasmine and *lingzhi* by the sea,
wondering how far away they really were,
whether they were shining still at the source
by the time their light reached me ...

I was told that everyone was born with a star
which glowed or dimmed with the fortunes of each.
I also heard people destined to be close
were at first fragments of the same star

and from birth went searching for each other.
Such parting, seeking, reuniting might take
three lifetimes with centuries in between.
I had thought all these were but myths ...

Now decades later, I read about the life of stars,
how their cores burn for ten billion years,
how towards the end, just before oblivion,
they atomize into nebulae of fragile brilliance –

ultra violet, infra red, luminous white, neon green or blue,
astronomical butterflies of gaseous light
afloat in a last waltz choreographed by relativity,
scattering their heated ashes into the void of the universe ...

Some of this cosmic dust falls onto our little earth
carrying hydrocarbon compounds, organic matter
able to mutate into plant and animal life,
a spectrum of elemental fragrances . . .

Perhaps on the dust emanating from one ancient star
were borne the first molecules of a *pandan* leaf,
a sprig of mint or basil, a vanilla pod, a vine tomato,
a morning frangipani, an evening rose, a lily of the night . . .

Perhaps our parents or grandparents or ancestors further back
strolling through a garden or a field had breathed in the scents
effusing from some of these plants born of the same star
and passed them on as DNA in the genes of which we were made . . .

Could that be why, on our early encounters, we already sensed
in each other a whiff of something familiar, why when we are near,
there is in the air some spark which seems to have always been there,
prompting us to connect our pasts, share our stories even as they evolve . . .

. . . till the day when we too burn away into dust
and the aromas of our essence dissipate
into the same kaleidoscope of ether light
to be drawn into solar space by astral winds . . .

. . . perhaps to make vanilla in a star to be
before the next lifetime of three?

Chapter 1
Where Do We Come From?

How did life originate on Earth? Was it the result of supernatural creation? Or are we the product of deliberate planting by advanced extraterrestrial civilizations? If life is the result of divine intervention, did life appear suddenly with all its functions and capabilities, or had the diverse forms of life on Earth developed over time from certain holy seeds? If extraterrestrials are involved, are we a duplicate of their forms, or were we created as an experiment? If so, did they actually visit Earth or did they deliver their experimental ingredients programmed with specific instructions to this planet by a space probe? Alternatively, maybe we were products of accidental developments, arising naturally without design. If so, what was the initial mix of ingredients? How complicated were the ingredients? How did these ingredients get to the surface of Earth? Were they present when the primordial Earth was formed, or could they have been brought here after the formation of Earth? Could these externally delivered ingredients include primitive life forms such as bacteria?

These are very ambitious questions which until recently would have been regarded as outside the realms of science. However, from the 1970s, we have witnessed the emergence of new scientific disciplines of astrochemistry and astrobiology. These new disciplines have opened new avenues to tackle the old question of the origin of life. Instead of speculation, conjecture, or faith, we can now attempt to answer this question in a scientific manner.

The oldest hypothesis, and also the most common among all cultures, is that life is the result of supernatural intervention. Most primitive cultures believe that they owe their existence to a supreme being. This theory, in its most general form, is impossible to refute by scientific method although specific theories with definite descriptions of sequence of events and the nature of the creation can be subjected to scientific tests.

Our Solar System resides in the Milky Way Galaxy, which has over 100 billion stars, many similar to our own Sun. The Universe as a whole has more than 100 billion galaxies similar to the Milky Way. The age of our Galaxy is estimated to be about 10 billion years old, and the Universe is only slightly older (currently believed to be about 14 billion years). Recent advances in planet detection

S. Kwok, *Stardust*, Astronomers' Universe, DOI 10.1007/978-3-642-32802-2_1,
© Springer-Verlag Berlin Heidelberg 2013

techniques have revealed over 700 planets around nearby stars. It is quite likely that planetary systems are extremely common around Sun-like stars. If we extrapolate the planet detection rate to distant stars, then the number of planets in our Galaxy could also run into hundreds of billions. Of course, we don't know what fraction of these planets harbors life as the Earth is the only place we know to possess life. But if life forms do exist elsewhere, then many would be inhabiting planets around stars that have been around much longer. Their civilizations would be millions, or even billions of years older than ours. Given the fact that human civilization only started thousands of years ago, and our technological societies only began hundreds of years ago, it is extremely likely that there are many alien civilizations that are much, much more advanced than ours. If this is the case, then the chance is high that some of them would have visited us already.

However, even if extraterrestrial life forms had visited us we may not have recognized them. For example, if our young and relatively backward technological society had the ability to go back several hundred years to leave behind a DVD containing thousands of pictures and videos and music, our ancestors would not be able to see it as more than a piece of shining metal, nor would they be able to decipher its contents. An artifact left behind by an alien advanced civilization is likely far too elusive or mysterious for us to notice or to comprehend. If extraterrestrial intelligent beings had visited the Earth, they would not have left primitive objects such as the pyramids or simple marks on the ground. The absence of evidence for visits by extraterrestrials is therefore no proof of their not having done so. If we were indeed visited, either by advanced life forms or by robots they sent, they could have easily seeded life on Earth without our ever realizing it had happened.

It is clear that some hypotheses on the origin of life, although within the realm of possibility, are difficult or impossible to disprove. As scientists all we can do is to use our present knowledge of astronomy, physics, chemistry, and biology to investigate whether theories of the origin of life stand up to observational and experimental tests.

The hypothesis of spontaneous creation, which states that life arises from nonliving matter, has a long history. The Greeks, for example, promoted the theory that everything is created from primary substances such as earth, water, air, and fire. The idea that plants, worms, and insects can spontaneously emerge from mud and decaying meat was popular up to the seventeenth century. This theory was put to severe tests in the seventeenth century when the Italian physician Francesco Redi (1626–1698) noticed that maggots in meat come from eggs deposited by flies. When he covered the meat by a cloth, maggots never developed. This experiment therefore cast doubts on the premise that worms originate spontaneously from decaying meat.

The invention of the microscope has revealed the existence of large varieties of microorganisms which are invisible to the naked eye. A Dutchman, Antonie van Leeuwenhoek (1632–1723), found microorganisms in water and therefore showed that minute life is common. Van Leeuwenhoek was a tradesman who lived in Delft, Holland and had no formal training in science. He did have good skills in grinding lenses and made a large number of magnifying glasses for observations. He had put everything imaginable under his home-made microscope. The list of samples that he had observed include different sources of water, animal and plant tissues, minerals, fossils, tooth plaque, sperm, blood, etc. By using proper lighting during his observations, he was able to see things that no one had seen before. Among his many discoveries, the most notable is the discovery of bacteria, tiny living, moving organisms that are present in a variety of environments. For his achievements, this amateur scientist was elected as a member of the Royal Society in 1680.

Van Leeuwenhoek believed that these life forms originate from seeds or "germs" that are present everywhere. A revised form of spontaneous creation therefore contends that while large life forms such as animals may have come from eggs, small microscopic creatures can still be created from the non-living. This question was finally settled by Louis Pasteur (1822–1895) who showed that the emergence of microorganisms is due to contamination by air. His pioneering experiment is the beginning of our modern belief that life only comes from life on Earth today.

If this is the case, then when did the first life on Earth begin and how? By the late nineteenth century, scientists realized that the Earth is not thousands or millions, but billions of years old. Although life can no longer be created in the current terrestrial setting, may be it was possible a long time ago when the Earth's environment was very different. With suitable mixing of simple inorganic molecules in a primordial soup, placed in a hospitable environment and subjected to injection of energy from an external source, life may have originated over a long period of time. Given the old age of the Earth, time is no longer an issue. The idea that the origin of life on early Earth could be explained using only laws of physics and chemistry was promoted by Soviet biochemist Aleksandr Ivanovich Oparin (1894–1980) and British geneticist John Burdon Sanderson Haldane (1892–1964) in the 1920s.

Their ideas were motivated by the success of laboratory synthesis of organics in the nineteenth century. Historically, the term "organics" was used to refer to matter that is related to life, which is distinguished from "inorganic" matter such as rocks. It was assumed that inorganic matter can be synthesized from the basic elements (such as atoms), whereas organic matter possesses a special ingredient called the "vital force". The concept that the "living" is totally separated from the "non-living" was entrenched in ancient view of Nature. To draw an analogy, the concept of "vitality" separating living from nonliving is equivalent to the concept of "soul" which supposedly distinguishes humans from other animals. The concept of vitalism can be summarized in the words of the nineteenth century physician–chemist William Prout (1785–1850): "(there exists) in all living organized bodies some power or agency, whose operation is altogether different from the operation of the common agencies of matter, and on which the peculiarities of organized bodies depend". As for the form of this "power", he said "independent existing vital principles or 'agents,' superior to, and capable of controlling and directing, the forces operating in inorganic matters; on the presence and influence of which the phenomena of organization and of life depend". This was the prevailing view in the nineteenth century.

The concept of "vitality" originated from simple observations that living things can grow, change and move, whereas non-living things cannot. These activities are now explained by the modern concept of "energy", which explains movement as the conversion from one form of energy (chemical) to another (kinetic). In spite of the introduction of the concept of energy, "vital force" remained a popular concept in chemistry. However, the physical form of "vitality" was never precisely defined nor quantified, although by the nineteenth century, it was believed to be electrical in nature.[1] Nevertheless, "vital force" was thought to be real as it was the absence of "vital force" that was assumed to make it impossible to synthesize organics chemically from inorganics. In 1828, Friedrich Wöhler (1800–1882) synthesized urea, an organic compound isolated from urine, by heating an inorganic salt ammonium cyanate. This was followed by the laboratory synthesis of the amino acid alanine from a mixture of acetaldehyde, ammonia, and hydrogen cyanide by Adolph Strecker (1822–1871) in 1850, and the synthesis of sugars from formaldehyde by Aleksandr Mikhailovich Butlerov (1828–1886) in 1861. While it was thought that a vital force in living yeast cells is responsible for the process of changing sugar into alcohol, Eduard Büchner (1860–1917) showed in 1897 that yeast extracts can do the same without the benefit of living cells. The successes of these artificial syntheses led to the demise of the "vital force" concept.

The discipline of biochemistry emerged from this philosophical change. Biochemistry is based on the premise that biological forms and functions can be completely explained by chemical structures and reactions. The catalysts that accelerate chemical reactions in biological systems are biomolecules that we now call enzymes. In 1926, James Sumner found that an enzyme that catalyzes urea into carbon dioxide (CO_2) and ammonia (NH_3) belongs to the class of molecules called proteins. James Batcheller Sumner (1887–1955) had only one arm, having lost the other due to a hunting accident when he was a boy. When he tried to undertake Ph.D. research in chemistry at the Harvard Medical School, he was advised by the chairman of the biochemistry department that he should consider law school as "a one armed man could never make it in chemistry". However, he did finish his Ph.D. at Harvard and took up a position as assistant professor in the Department of Physiology and Biochemistry in the Ithaca Division of Cornell University Medical College. Although he had limited equipment or research support, he took on the ambitious project to isolate an enzyme. After 9 years, he crystallized the enzyme urease. His results were doubted by his contemporaries and his work was only fully accepted in 1946 when he was awarded the Nobel Prize.

Many other digestive enzymes also turned out to be proteins. The magic of life has therefore been reduced to rules of chemistry. By the early twentieth century, this has become the new religion in science. Living matter, although highly complex, is nothing but a large collection of molecules and the working of life is

---

[1] It is interesting that the quantification of "soul" can be found in modern popular culture. The 2003 movie "21 Grams" mentions the supposed scientific study showing that people lose 21 g in weight at the time of death, presumably due to the separation of soul from the body.

no more than a machine having numerous molecular components working with each other. Under such a belief, the origin of life could also be understood through a set of chemical reactions. These new laboratory developments therefore set the stage for the adaptation of the Oparin-Haldane hypothesis as the dominant theory of the origin of life by the mid-twentieth century.

Although the Oparin-Haldane hypothesis had a sound scientific basis, it was also politically convenient for Oparin because the idea of life originating from non-living matter fits in well with the Marxist philosophical ideology of dialectic materialism. Oparin graduated from Moscow University in 1917, right at the time of the Russian revolution. He began his research in plant physiology and rose to become the director of the Institute of Biochemistry of the USSR Academy of the Sciences in 1946. Beginning as early as 1924, he explored the idea that life could originate from simple ingredients in the primitive Earth. Oparin was very successful in the Soviet Union, becoming Hero of the Socialist Labor in 1969, recipient of the Lenin Prize in 1974, and five Orders of Lenin. It is interesting that Haldane, a British geneticist, was also a devout Marxist. He was a member of the communist party of Great Britain, although in his later years he broke away from Stalinism because the Soviet regime was persecuting scientists in the Soviet Union. In 1956, Haldene left his position at University College London and moved to India, as he disagreed with the British world political stand on the Suez Canal at that time. He became a vegetarian and wore Indian clothing. He died in India in 1964.

It is difficult to know whether the Marxist philosophical leanings of Oparin and Haldane had any bearings on their independently developed ideas on the origin of life, but it is probably fair to say that their theory had more in common with a mechanical view of the universe than a spiritual one, as was popular at the time. Oparin's work was not known in the west until the translation of his book "*The Origin of Life*" into English in 1938 and republication in the U.S. in 1952, and Haldane's ideas were dismissed as mere speculations. Haldane wrote many books, some of them popular ones, even some for children. The fact that he was a prolific and eloquent writer certainly helped to keep him in the public limelight; otherwise his work on the origin of life might have been forgotten.

The Oparin-Haldane hypothesis only gained respectability after the experimental demonstration in the 1950s. In a milestone experiment in 1953, Stanley Miller (1930–2007) and Harold Urey (1893–1981) of the University of Chicago showed that given a hospitable environment (e.g. oceans) and an energy source (e.g. lightning), complex organic molecules can be created naturally from a mixture of methane, hydrogen, water, and ammonia. Using a flask to simulate the primitive atmosphere and ocean and injecting energy into the flask by electric discharge, Stanley Miller found that a variety of organic compounds such as sugars and amino acids emerged in this solution. This experiment had an extraordinary impact on the thinking of the scientific community. For the first time, spontaneous creation seemed to be a possibility (Fig. 1.1).

Stanley Miller was a graduate student at the University of Chicago, originally working with the nuclear physicist Edward Teller. After Teller left Chicago, Miller had to find a new advisor and he approached the geochemist Harold Urey, who had

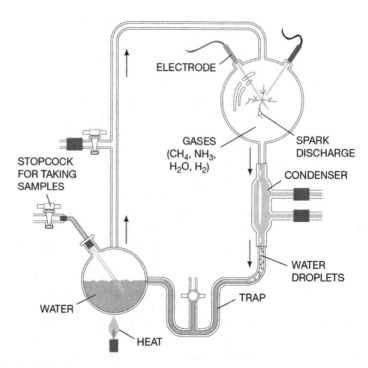

Fig. 1.1 The Miller–Urey experiment.
The experiment consists of a simple flask (*upper right*) containing a mixture of methane, ammonia, water and hydrogen. An electric spark is introduced. The chemical reaction products collected include amino acids and other complex organics, showing that biomolecules can be synthesized naturally under conditions of the early Earth

suggested that the atmosphere of the early Earth had a composition made up of water, ammonia, and methane but no oxygen. Miller wanted to test what kind of chemistry could be at work under conditions of the early Earth. Urey had thought that for interesting results to emerge, the experiment had to run a very long time. Everyone was surprised when Miller observed the presence of amino acids in the flask after only a few days. The experiment was reported in the journal *Science*. Legend has it that although Miller initially put Urey's name on the paper, Urey declined citing that "I already have a Nobel Prize" and left his student to take full credit for the discovery. The Miller–Urey experiment was described by Carl Sagan as "the single most significant step in convincing many scientists that life is likely to be abundant in the cosmos." For the second half of the twentieth century, the theory of life emerging spontaneously from simple molecules in a primordial soup in the young Earth became widely accepted by the scientific community.

Another theory of the origin of life considers the possibility that life is common everywhere in the Universe and is spread from place to place. The hypothesis of panspermia stipulates that life on Earth originated from outside and was delivered to Earth. More than 2,000 years ago, the Greek philosopher Anaxagoras (~500

BC–428 BC), who discovered the nature of eclipses, had already outlined the principle of panspermia. He considered that the seeds of life are already in the Universe, and they will take root whenever the conditions become favorable.

Back in 1871, the German physiologist Hermann von Helmholtz (1821–1894) wrote that "who could say whether the comets and meteors which swam everywhere through space, may not scatter germs wherever a new world has reached the stage in which it is a suitable place for organic beings." The Swedish chemist Svante Arrhenius (1859–1927) promoted in his book *"Worlds in the Making"* in 1904 (1 year after he won the Nobel Prize—the first Swede to receive the honor) the idea that simple life forms (e.g. bacteria) spread from star to star by long journeys through the interstellar medium. His idea was that small particles containing seeds of life could be propelled between planetary systems by radiation pressure, the force that light exerts on solid bodies. He believed that these spores frozen in the low temperature of interstellar space could be revived again once they reach favorable surroundings after journeys of thousands of years. However, there was no empirical evidence at the time for the mechanisms that he was considering and interest in panspermia died down in the 1920s.

While the external hypothesis does not solve the problem of the origin of life, but simply shifts the problem to somewhere else, it cannot be dismissed easily. It is quite possible that there are locations with more favorable conditions for the creation of life and we are the beneficiaries. The most widely promoted hypothesis of life arriving from space in recent times is in the works of Fred Hoyle (1915–2001) and Chandra Wickramsinghe. These authors argue that if life can develop from inorganic matter from Earth, life must have been common in millions of other solar systems as the Galaxy is old (~10 billion years). Living organisms from those systems could just as easily have been transported to our Solar System and seeded life on Earth. They also cited the fact that microorganisms can survive and indeed thrive under extreme conditions as evidence that bacteria can endure long interstellar journeys. The analogs of bacteria revived from bees embedded in amber for 25–40 million years and in 250 million year old salt crystals have also been cited as evidence of the viability of panspermia.

If the Oparin-Haldane hypothesis is correct that life on Earth originated from simple inorganic molecules, then similar processes could also be at work elsewhere. This possibility was raised in the book *Life in the Universe* by Oparin and Soviet astronomer V. Fesenkov in 1956. Technological advances, in particular in the form of the space program in the U.S. and in the Soviet Union, heightened the hope that extraterrestrial life could become a subject for experimental studies. Probes and landers to the Moon and Mars could search for signs of life. The first serious attempts to address this question were the two *Viking* spacecrafts which landed on the surface of Mars in July and September of 1976. The *Viking missions* were equipped with biological experiments to search for signs of metabolic activities as signs of life. While the experiments found that the surface of Mars was chemically active, there were no definite indications of biological activity. By the end of the mission, scientists came to the reluctant conclusion that extraterrestrial life has not been found on Mars.

As of 2012, there has been no empirical evidence for the existence of extraterrestrial life forms such as bacteria anywhere in the Solar System, or beyond. However, it has been known since the mid-nineteenth century that meteorites contain organic material. The Alais meteorite that fell in Alais, France, in 1806 and the Kaba meteorite that fell near Debrecen, Hungary, in April, 1857, were found to be rich in organics upon analysis. This was the first indication that complex organic materials may not be the sole domain of the Earth, and are actually present beyond the Earth, at least in the Solar System. In the nineteenth century, and as a matter of fact during most of the twentieth century, it was commonly believed that life on Earth was unique, and organic matter should only be found on Earth. The concept that organic matter resides in meteorites originating outside of the Earth did not take hold until the mid-twentieth century, although evidence for it had been around for over a century.

At the beginning of the twenty-first century, here is how we stand on the question of origin of life on Earth. On one side we have the chemical origin of life in the form of the Oparin-Haldane hypothesis and support from the Miller–Urey experiment. On the other side we have the biological delivery in the form of the theory of panspermia of Arrhenius and Hoyle and Wickramsinghe. Is there a middle ground? We now know that organic matter is not only present in the Solar System, but elsewhere in the Universe as well. In this book, we will tell the story of how we come to learn that organic matter is prevalent throughout the Universe. We now know that stars can make large quantities of organic compounds efficiently. These organics are contained in stardust, tiny specks of solids manufactured by stars. We have found that such stardust particles are made in the last one million years of a star's life, and they are spread throughout the Milky Way Galaxy. After a long journey through space, they became part of our early Solar System, and we now have direct evidence of their presence in our midst. We will describe how we learned about the existence of organic matter in the Universe, how we discovered that stars are capable of producing organics, and how these stellar materials might have had an effect on the origin of life on Earth.

A brief summary of this chapter
How we learned about organic matter in universe, about stars producing organics, how they might affect origin of life on Earth.

Key words and concepts in this chapter
- History of hypotheses on the origin of life
- Supernatural, extraterrestrial intervention, spontaneous creation
- Vital force as a component of organic matter
- Biological forms and functions can be explained by biochemistry
- Oparin-Haldane hypothesis for a chemical origin of life
- The Miller–Urey experiment as a simulation of chemical processes leading to life

- Panspermia
- Organic matter in the Universe

Questions to think about
1. Even as early as 4,000 years ago, ancient people already pondered about the question of the beginning of humans. Why do you think humans had the need and urge to seek an answer to this question?
2. What do you think of the concept of "vitality"? Is it reasonable to think that the living and the non-living are distinguished by something significant?
3. Energy is also an abstract entity that we cannot touch or feel. Is it more real than vitality?
4. What do you think of the field of biochemistry on philosophical grounds? Is it a reasonable assumption that biology can be reduced to chemistry?
5. Why is the Miller–Urey experiment significant? Why didn't people think of doing this before?
6. Do you think that life is unique? As of 2012, there is no evidence for the existence of extraterrestrial life. Do you think that there is life beyond the Earth?

Chapter 2
Rocks and Dust in the Planetary Neighborhood

The planet we live on, the Earth, is a chunk of rock partially covered with liquid water and overlaid with a thin blanket of gaseous atmosphere. Liquid oceans and polar ice caps cover three quarters of the Earth's surface. The continents, on which we walk and build our cities and villages, are made up of rocks. However, the rocky crust of the Earth is not limited to the continents, but extends to the ocean floors. These rocks are aggregates of minerals, which are solid-state compounds of common elements such as oxygen, silicon, aluminum, iron, calcium, sodium, potassium, and magnesium. Three of our Solar System neighbors, Mercury, Venus, and Mars, have similar rocky surfaces and the four together are collectively known as the "terrestrial planets". The rocky nature of Mars is most vividly illustrated by the landscape images sent back by the Martian rovers *Spirit*, *Opportunity*, and *Curiosity*. In contrast, the other four planets in the outer Solar System—Jupiter, Saturn, Uranus, and Neptune —are gaseous in nature and do not possess a solid surface. The only anomaly is Pluto, the outermost member, which is believed to be made up of water ice and is no longer considered a planet.

Most planets have moons that revolve around them. Our own Moon, the only natural satellite of the Earth, is also rocky. When Galileo Galilei (1564–1642) observed the Moon with a telescope in 1610, he did not find a perfect, smooth celestial body, but instead an uneven and rough surface. The majority of the topographical features of the Moon turned out to be craters, or scars left over from external impact events. The rocky nature of the Moon is clearly illustrated by the Apollo astronauts who walked and drove vehicles on the surface of the Moon. Other planetary satellites, such as the Martian moons Phobos and Deimos, are also rocky. So are the moons of Jupiter such as Io, Europa, Ganymede, and Callisto. So is the largest moon of Saturn–Titan. Pictures brought back by the European Space Agency's *Huygens probe* showed the rocky nature of Titan's surface most clearly and dramatically (Fig. 2.1).

On a smaller scale, there are the asteroids. Asteroids are small, rocky objects that revolve around the Sun. The largest asteroid known, Ceres, has a size of 940 km and a mass 10,000th that of the Earth. Many asteroids are concentrated in the "asteroid belt" between the orbits of Mars and Jupiter. The number of asteroids known

S. Kwok, *Stardust*, Astronomers' Universe, DOI 10.1007/978-3-642-32802-2_2,
© Springer-Verlag Berlin Heidelberg 2013

Fig. 2.1 A view of Titan's surface taken by the *Huygens probe*.
A view of the surface of Titan as taken by the *Huygens probe* during its fall through Titan's atmosphere after its release from the *Cassini* spacecraft on January 14, 2005. Photo credit: ESA

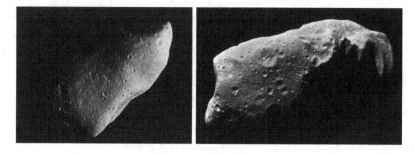

Fig. 2.2 Images of the asteroids Gaspra and Ida.
These images of Gaspra (*left*) and Ida (*right*) were taken by the *Galileo* spacecraft. Marks left by past impacts can clearly be seen on the surface. The longest dimension is about 20 km for Gaspra and 50 km for Ida. Photo credit: NASA/JPL

exceeds 100,000, a number likely to increase rapidly as larger telescopes are put into action for their search. Close-up photographs taken by spacecrafts have revealed that asteroids are irregular objects. When the *Galileo* spacecraft flew by the asteroid 951 Gaspra on October 29, 1991, the pictures of the asteroid showed that its surface is marred by deep scars created by a long history of impact events. The images of Gaspra and Ida (Fig. 2.2) and of Vesta (Fig. 2.3) definitely carry home the message that these heavenly bodies look very much like ordinary rocks.

If asteroids are rocky, then it would be possible to land on them. Indeed a Japanese space mission did exactly that. The *Hayabusa mission* was launched on May 9, 2003 and reached asteroid 25143 Itokawa in September 2005. It descended on the asteroid in November 2005 and collected samples from the surface of the

Fig. 2.3 Image of the asteroid Vesta.
This picture of Vesta was taken by the *Dawn* spacecraft. *Dawn* was launched in September 2007 and reached the asteroid in July 2011 after an almost 4 year journey across the Solar System. At a distance of 41,000 km, the surface of Vesta can be clearly seen to have a rocky appearance. The main difference between asteroids and the Earth and the Moon is that they are not necessarily spherical in shape. However, Vesta is nearly spherical with a diameter of about 500 km. Photo credit: NASA/JPL

asteroid before returning to Earth on June 13, 2010. A capsule containing the rock sample was released from the spacecraft and landed safely in Australia. Although the amount of asteroid materials contained in the return sample was small, it did allow scientists a direct look into the chemical composition of the asteroid and establish the asteroid origin of some of the meteorites collected on Earth.

How small can an asteroid be? There is no clear definition on the minimum size of an asteroid. We do know that solid debris of all sizes are present in the interplanetary space. The commonly used dividing line is a size of 100 m, although this is entirely arbitrary. Any object larger than 100 m is called an asteroid, smaller a "meteoroid", which we define as small asteroids with orbits around the Sun that cross the Earth's orbit. When a meteoroid enters the Earth's atmosphere, the friction of its fall heats up the surrounding air, leading to a streak of light across the sky which we see as a meteor (commonly known as a "shooting star"). Although now we can predict the coming of meteor showers, we still cannot predict the exact occurrence of an individual meteor. Meteors are examples of the changeable nature of the heavens. Although the Sun, the Moon, the stars, and the planets all have regular, predictable movement patterns in the sky, the coming of meteors cannot be predicted. Such unpredictable events in the heavens have been used for astrological purposes. For example, the occurrence of a meteor was interpreted in China as a sign for the fall of a major figure, a court official, a commanding general.

On a typical clear moonless night, several meteors can be observed in an hour across the sky. These are called sporadic meteors as their appearance can be sudden and unpredictable, and they seem to appear in random directions in the sky. These are in contrast to shower meteors, which seem to originate from a fixed point in the sky and can be expected to occur at certain times of the year. For example, a meteor shower that occurs every August seems to radiate from the constellation of Perseus and is named the Perseid meteor shower. In 1833, a spectacular meteor shower was

Fig. 2.4 Engraving showing
the 1833 Leonid meteor
showers.
Meteor showers can be
dramatic events as seen in this
engraving showing the 1833
Leonid meteor showers

seen to come from the constellation of Leo (Fig. 2.4). It has been said that the great
meteor shower of 1833 was responsible for the ignition of public interest in
astronomical research in the USA. This meteor shower, named the Leonid meteor
shower, had a repeat performance on 1866. In fact, the historical record shows that
this same meteor shower has been observed as early as 899 AD. We now know that
the periodic nature of meteor showers is related to earth-orbit crossing comets
(Chap. 18).

As a meteoroid passes through the atmosphere it burns itself up due to atmospheric friction and gives off visible light in the process. It may also fragment into
multiple pieces. Sometimes a meteoroid will be completely vaporized in its passage
through the atmosphere. In fact, most meteoroids lose most of their mass. If a
fraction of the meteoroid survives at all through the atmosphere, its remnant on the
ground is called a meteorite.

Falls of meteorites have been recorded throughout history, dating back over
1,000 years. The oldest meteorite fall on record is the Nogata meteorite recovered in
Japan on May 19, 861 A.D. The earliest European record was that of the Ensisheim
meteorite in 1492. The meteorite that fell on June 16, 1794 was at a location near a
major European city, Siena, Italy, and was witnessed by many people. In spite of the
numerous reports of meteorite falls in Europe, they were not taken seriously. In
China, Japan, and Korea, imperial and provincial records contained hundreds of

reported cases of meteorite falls, but there was little interest in seeking out the origin of these events. In Europe, the more religious considered meteorites to be acts of the devil. Those who were brave or imaginative enough to suggest otherwise were condemned as heretics. There was a strong belief that the heavens are divine, holy, and unchanging. Since the time of Aristotle, the celestial sphere was seen to be separate and detached from the terrestrial earth, where hurricanes, floods, volcanos, earthquakes, and other unpredictable misfortunes happen. In religious circles, such calamities were commonly believed to be caused by sins of men. This is in stark contrast to the heavens which are peaceful and everlasting. There was a strong reluctance to associate temporal phenomena such as meteors with anything in the celestial sphere. Mud and rocks are supposed to be confined to the domains of the Earth, and could not have descended from the heavens.

Even in modern times, meteorites were thought to be created in the atmosphere by lightning, or by accretion of volcanic dust. One of the most famous quotes was that of Thomas Jefferson (1743–1826), who reacted to the reported 1807 fall of a meteorite in Connecticut with the comment "it is easier to believe that two Yankee professors would lie than that stones should fall from heaven". In 1794, a German scientist, Ernst Friederick Chladni (1756–1827), suggested that meteorites are the result of fireballs, and meteorites do not originate from the clouds, but from outside of the Earth. His idea received no support from his contemporary scholars. In 1802, an English chemist, Edward Charles Howard (1774–1816), analyzed four meteorites and found their chemical composition to be similar but different from terrestrial rocks. It was the chemical analyses and microscopic examinations of meteorites that finally persuaded the scientific community of the extraterrestrial origin of meteorites and that did not happen until the early nineteenth century.

Before 1969, there were only 2100 known meteorites catalogued. The remarkable increase in the recovery of meteorites in recent years is due to the activities in Antarctica. Although meteorites fall randomly all over the world, the white, icy background of Antarctica makes it much easier to spot one. Also since the rocky surface of Antarctica is hidden below 3,000 m of ice, any piece of rock one sees on the surface is likely to be extraterrestrial in origin. A meteorite fallen on ice is also less likely to have been damaged and is better preserved through time. The first meteorite in Antarctica was found in 1912, but in 1969, the Japanese Antarctic Research Expedition found 9 meteorites near Yamato Mountains and realized that the movement of ice sheets can bring meteorites close together in concentrated groups. Since then there have been many search programs by different countries. For example, the American Antarctic Search for Meteorites (ANSMET) team has recovered over 20,000 meteorites. As of 2012, there are over 45,000 known meteorites. Other promising recovery sites are deserts, where the barren landscape also allows for easy identification. Successful recoveries have been made in deserts in Chile, Namibia, and Australia.

One good thing about meteorites is that we can pick them up and examine them. An initial impression of meteorites is that they resemble rocks. This in itself is significant, as meteorites are extraterrestrial objects; we have direct, first-hand experience that there are rocks in the sky. However, they are on average heavier than a rock of similar size on Earth.

Fig. 2.5 Micrometeorites
collected in Antarctica.
This sample of
micrometeorites was
collected by a French–Italian
expedition to Antarctica in
2003. Image credit: CNRS
Photothèque, Michel
Maurette

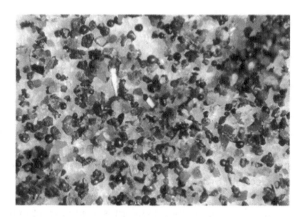

Although we generally associate meteoroids with macroscopic objects of meters
or centimeters in size, the remnants of small meteoroids (those of golf balls or
marbles sizes) may leave too small a remnant to be seen by the eye. A very small
meteoroid (with sizes smaller than a millimeter) may even float through the
atmosphere without burning. The remnants of these meteoroids are called
micrometeorites (Fig. 2.5). Samples of these particles have been collected in
Greenland and Antarctica where they are easier to spot on an ice/snow surface.
From the collected samples, it is estimated every year 40,000 tons of
micrometeorites land on the Earth.

Our current thinking of the origin of micrometeorites is that they are remnants of
comets. Since micrometeorites come from space, they must be flying around in
interplanetary space. As solid objects, the terrestrial planets, moons, and asteroids
can reflect sunlight and can be seen as bright visual objects in the sky. Smaller
solids like meteoroids can also be seen as meteors as they burn through the
atmosphere.

Although meteorites can be found by just random searching of the ground, they
are much easier to find if we know that one is falling. The fall of a large meteorite is
often accompanied by bright fireballs and loud sonic booms. On average, two or
three large fireballs are observed worldwide per night. In a way, a fireball is just a
very bright meteor (Fig. 2.6). The International Astronomical Union defines a
fireball as a meteor that is brighter than any of the planets. These falls often create
quite an alarm among the populations that witnessed them. Nowadays, the sky is
systematically watched for fireballs through the installation of a system of wide-
angle cameras. By comparing the images of the fireballs in different cameras, the
location of the fall can be identified, therefore increasing the chance of recovering
the meteorite.

One of the earliest records of tracking the fall of a meteorite is the meteorite that
landed near Orgueil in southern France on May 14, 1864. A brilliant fireball
accompanied by loud explosions was seen and heard all over southern France
shortly after 8 PM. Most of the stones fell near the village of Orgueil and over 20
pieces of black rock were immediately collected by local villagers. The largest

Fig. 2.6 Fireballs.
This picture of a fireball was taken on September 30, 2008 at the Black Mesa State Park in Oklahoma, USA by Howard Edin. This fireball meteor appeared about 2 AM and can be seen passing through the constellations of Taurus (*top*) and Orion. Photo credit: Howard Edin

Fig. 2.7 Carbonaceous chondrite meteorites. Carbonaceous chondrite meteorites such as the Allende shown in this picture are often black or dark gray in color, and are rich in carbon. These are the kinds of meteorites in which extraterrestrial organics are found. Photo credit: Toby Smith

piece is as large as a man's head, but typically the collected pieces are fist-sized objects. A total of 11.5 kg of the meteorite remnants were recovered, with the largest piece (9.3 kg) now residing in the Paris Museum. Between 1864 and 1894, 47 pieces of scholarly work were published on the meteorite.

As times moved on, scientists found that one of the distinguishing features of meteorites is their texture. Inside the meteorites are millimeter-size ellipsoidal structures called chondrules. The shapes and structures of the chondrules led us to believe that these chondrules were formed in a molten state. One special kind (about 4 % of all the meteorites that fall to the Earth) is called carbonaceous chondrites (Fig. 2.7). They are among the most primitive objects in the Solar System.

In 1834, a Swedish chemist, Jöns Jacob Berzelius (1779–1848), was the first to discover that meteorites contain organic materials. Berzelius was considered the

Fig. 2.8 Murchison
meteorite.
The meteorite that fell near
the town of Murchison,
Australia, is one of the
carbonaceous meteorites that
are found to contain a large
amount of organic material

father of Swedish chemistry and his proposed system of chemical notation using a combination of letters and numbers to signify elements and their proportions (e.g., H_2O) in a compound is still in use today. Confirmation of this discovery had to wait over a hundred years. In March, 1961, Bartholomew Nagy and Douglas Hennessy of Fordham University in New York, together with a petroleum chemist, Warren Meinschein, from the ESSO Research Corporation, reported in a meeting of the New York Academy of Sciences that they found hydrocarbons in the Orgueil meteorite that resemble products of life. Nagy and his team performed their analysis on a sample obtained from the Museum of Natural History in New York. They took incredible care to avoid the possibility of contamination. Using a mass spectrometer, they identified various hydrocarbons including paraffins similar to those found in animal products such as butter. From this study, Nagy concluded that these organic compounds are indications of biogenic activities beyond the Earth.

In a paper published 3 weeks after the New York meeting, John Desmond Bernal (1901–1971) of the University of London commented on Nagy's report and discussed the possible origin of the organics. If the hydrocarbons are the products of life, then they must have been inherited from a planetary body, from which the Orgueil meteorite originated. However, most bodies in the asteroid belt are believed to be too small and too dry to harbor life. He even speculated that the Orgueil meteorite could have been sent from the primitive Earth, and arrived back at the Earth after a journey lasting millions of years in interplanetary space. The possibility that the meteorite came from an Earth-like planet in another solar system was also mentioned. In any case, the presence of organics in meteorite is difficult to explain.

On September 28, 1969, half a ton of fragments of a meteorite fell near the town of Murchison in Victoria, Australia. Following a bright fireball and a loud boom, fragments of the fallen meteorite were found over an area larger than 13 square km. Individual pieces as large as 7 kg were recovered. A fragment of the Murchison meteorite is shown in Fig. 2.8. Analysis of the Murchison meteorite by John Cronin and Sandra Pizzarello of Arizona State University has shown they contain complex organic compounds with both aromatic (ring-like) and aliphatic (chain-like) structures. The term "aromatics" was first used in the mid-nineteenth century to refer to chemical substances with notable aromas, but now the term is used to refer to a class of molecules consisting of rings of six carbon atoms. The simplest example is benzene, which consists of six carbon atoms arranged in the form of a ring with six hydrogen atoms each attaching to one carbon atom. In contrast, aliphatic compounds consist of carbon atoms arranged in the form of a chain. Both aromatic and aliphatic compounds form the basis of molecules of life. Many common biochemical substances, including amino acids, fatty acids, purines, pyrimidines, and sugars are found in the Murchison meteorite.

One of the people who has contributed much to our modern understanding of organics in meteorites is Sandra Pizzarello. She was originally from Venice, Italy and already had four children before she moved to the U.S. with her husband. When her youngest child went to grade school, she decided to go back to university to study biochemistry. One of her teachers was a young professor named John Cronin. From that point on, Pizzarello and Cronin carried on decades of collaboration, discovering and analyzing organics in meteorites.

On the morning of January 18, 2000, a fireball appeared in the sky of the Yukon Territory of Canada, and the colorful event lasting over 10 min was witnessed by hundreds of observers. However, subsequent aerial flights were unable to find a crater or a site of impact of the meteorite. Fortunately, fragments were found within 1 week on the frozen surface of Tagish Lake. This is a big break for scientists as the frozen lake was perfect for keeping the meteorite in pristine condition.

I consider meteorites gifts from heaven, as they give us the direct and complete access to extraterrestrial material. We can examine meteorites by visual inspection, subject them to passive observations by instruments, as well as perform active manipulations by experiments. This is far superior to the limited ability of remote astronomical observations of celestial objects. We can measure and determine their physical and chemical properties, and derive the abundance of various chemical components as well as their isotopic ratios. Laboratory techniques that have been applied to meteorites include spectroscopy, nuclear magnetic resonance (NMR), X-ray absorption near-edge structure (XANES), and electron paramagnetic resonance (EPR).

If we go to the countryside away from the artificial light of the cities and step out about 1 h after sunset on a clear, moonless evening in the Spring, we can see a cone of diffuse light shaped like a pyramid spread out near the horizon with the brightest region near the point where the Sun has just set. This cone of light is called the zodiacal light (Fig. 2.9). The zodiacal light is best viewed from the tropics near the equator as it spreads out high in the sky. At high Northern or Southern latitudes, the zodiacal light lies closer to the horizon. At its best, the brightness of the zodiacal light can rival that of the Milky Way. However, the nature of the two is very different. Although it appears diffuse to the naked eye, the Milky Way is made up of light from billions of distant stars in our Galaxy (Fig. 2.10). The zodiacal light is much closer; it originates from the ecliptic, the plane in the Solar System where most of the orbits of the planets lie. If we observe the zodiacal light with a telescope, we will not be able to separate it into distinct stars. In fact, the zodiacal light originates from sunlight reflected by a large collection of tiny dust particles on the ecliptic plane. These particles are so small that we cannot see them with our eyes if we hold them in our hands. The particles that are responsible for the zodiacal light have sizes between a fraction of a micrometer (μm, or one thousandth of a millimeter) to several micrometers and they are called interplanetary dust particles.

Fig. 2.9 Zodiacal light.
This picture of the Zodiacal
Light was photographed by
Dominic Cantin near Quebec
City, Canada. The bright spot
near the horizon is the planet
Venus. Photo credit: Dominic
Cantin

Fig. 2.10 The Milky Way.
A panoramic view of the
Milky Way. The dark patches
in this picture are due to
absorption by interstellar dust
along the light of sight. The
center of the Milky Way
galaxy is located in the
constellation Sagittarius,
close to the border of
Scorpius and Ophiuchus.
Photo credit: Wei-Hao Wang

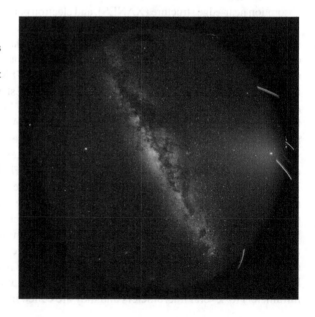

The origin of the dust particles that make up the zodiacal light is, however, unclear. They could come from fragments of either asteroids or comets. From an analysis of the radiative forces on these particles from solar radiation as well as gravitational forces exerted by the planets, Peter Jenniskens of the SETI Institute in Mountain View, California, came to the conclusion that they are most likely to be of cometary origin.

Our modern studies of interplanetary dust particles are not limited to the observation of reflected sunlight from these particles. In fact, we can go out and actively collect them in the upper atmosphere. By using high-flying aircraft, we can collect samples of interplanetary dust particles and bring them back to Earth for scientific analysis. We will talk more in detail about them in Chap. 16.

So we have solid rocks in the inner Solar System in the forms of asteroids and interplanetary dust particles, and they occasionally run into the Earth. The fall of asteroids through the atmosphere may leave visible trails in the form of meteors, marks on the Earth's crust, or remnants on the ground as meteorites. However, the Solar System does not end at the orbit of Pluto and astronomers now believe that there are many rocky objects belonging to the Solar System beyond the orbit of Pluto. Pluto is in fact only one of tens of thousands of objects in a thick ring around the Sun. This ring extends to 50 astronomical units (A.U., the distance between the Earth and the Sun) and is named the Kuiper Belt in honor of Gerard Peter Kuiper (1905–1973), who together with Kenneth Essex Edgeworth (1880–1972) predicted its existence. The Kuiper Belt is probably the remnants left over from the formation of the Solar System, where the inner part condensed into the major planets, leaving the thousands of small, icy objects stranded in the outer parts.

The first Kuiper Belt Object (KBO) was discovered in 1992. Since then nearly one thousand new members of the Solar System have been found, several of which are as big, or even bigger, than Pluto. These discoveries have raised doubts about whether Pluto can genuinely be called a planet, as it is no different in character from the other KBOs. The issue on the status of Pluto was debated by the International Astronomical Union in 2006 during its General Assembly in Prague. The General Assembly is held once every 3 years and is the occasion where astronomers from all over the world gather to disseminate the latest astronomical discoveries and to debate policy issues. As a result of a vote by the delegates present in Prague, Pluto was removed from the list of planets. This outcome was a surprise to the general public, whom for years have been told that there are nine planets but now all of a sudden found that there are only eight. This demotion of Pluto is not because astronomers have a change of heart but because new discoveries in our Solar System studies put the nature of Pluto in a new light.

We now believe there are over 100,000 KBOs with sizes larger than 100 km populating the Kuiper Belt. If we count smaller objects, the numbers are staggering. Extrapolating from the discovery of KBOs of sizes as small as 10 m in 2006 by Taiwanese astronomer Hsiang Kuang Chang, the population of such small bodies in the Kuiper Belt is estimated to be in the quadrillions (10^{15}).

So is the Kuiper Belt the outer limit of the Solar System? No, beyond the Kuiper Belt is an immense sphere called the Oort Cloud (named after Jan Hendrik Oort,

1900–1992) which extends from 3000 to 100,000 AU from the Sun. The outer edge of the Oort Cloud extends one third of the way to our nearest stellar neighbor Proxima Centauri. This Oort Cloud is the home of the long-period (>200 years) comets, which when perturbed by the passage of a nearby star, can venture periodically into the inner Solar System. While on this journey, the proximity to the Sun causes its icy surface to evaporate, giving it a beautiful tail that we admire. For the other dormant comets that reside in the Oort Cloud, the Sun is too far away and its radiation too weak to cause any harm. Current estimates put the number of comets in the Oort Cloud in the trillions, or even tens of trillions.

So the Earth and its surroundings have a lot of solid materials. The size of these solids ranges from less than a micrometer (in the form of interplanetary dust) to kilometers in the form of asteroids and planetary satellites. As the Earth moves around the Sun, it passes through this concentration of interplanetary dust particles and accretes about a hundred tons every day. Unknown to most of us, we absorb a large amount of invisible fine dust from space. Our Earth is therefore not a closed system as we are constantly bringing in extraterrestrial materials into our local environment. We can consider this a form of rain from heaven.

Where did these rocks come from? For one hundred years, scientists believed that the Solar System was created out of a collapsed interstellar cloud of dust and gas. Rotating together, these primordial gases are stirred and mixed in an oven-like condition. When these gases cooled, molten rocks condensed out of this primordial nebula, giving us the planets, asteroids, comets, meteoroids, and other minor solid bodies in the outer Solar System.

In this picture, every piece of solid in the Solar System was freshly made. All molecules and dust that it inherited from the parent interstellar cloud were completely destroyed. All the memories were erased and all connections to the past were lost. Is that really the case? Now astronomers are finding evidence that it may not be completely true. In the remaining chapters of this book, we will discuss the discovery of stardust, small solid particles made by stars. This discovery shows that solids are not the sole domain of the Earth and the Solar System, but in fact are present in stars and in the Galaxy as well. Due to the existence and prevalence of these solid particles in the general interstellar space, the Solar System may not be completely isolated from the rest of the Galaxy as previously thought. There is evidence that the early Solar System has been enriched/contaminated by stardust. The Solar System was not a completely new entity formed 4.6 billion years ago, but has inherited materials from other stars in the Galaxy. This star-Solar System connection is the theme that we will explore in this book.

A brief summary of this chapter
Earth and surroundings have a lot of solid materials. Earth is not a closed environment.

Key words and concepts in this chapter
- Different forms of solids in the Solar System
- Planets and planetary satellites
- Asteroids
- Meteoroids, meteors, meteorites, and micrometeorites
- Extraterrestrial origin of meteorites
- Organics in the Orgueil and Murchison meteorites
- Zodiacal light and interplanetary dust particles
- Kuiper Belt objects
- Rain from heaven

Questions to think about
1. Is it surprising that there can be rocks that fall from the sky? Put yourself in the minds of people 200 years ago and discuss why it was so difficult for people to accept the extraterrestrial origin of meteorites.
2. Meteorites provide us a link to the extraterrestrial world. What are our other links? Can you compare the effectiveness of these links in learning about the extraterrestrial world?
3. At the meeting of the New York Academy of Sciences in 1961, Nagy, Meinshein and Hennessy were quoted to state that "biogenic processes occur and that living forms exist in regions of the universe beyond the earth" (Bernal, J.D. 1961, *Nature*, **190**, 129). Do you think it is possible?
4. Since the beginning of the twentieth century, astronomers have been studying galaxies as far as billions of light years away. The Kuiper Belt is in our Solar System and is relatively nearby, but they were not known until the late twentieth century. Why did astronomers pursue objects so distant and at the same time were so ignorant about objects in our own backyard?
5. The Earth is subjected to a constant rain of external fine dust particles at a rate of a hundred tons a day. What are the effects of this rain? Should we be concerned about it?

Chapter 3
Impacts from Beyond

Geologists have long been familiar with volcanic eruptions and it is known that the eruption openings of volcanoes form circular depressions which are called craters. A large fallen meteoroid, say, one weighing a ton and over a meter in size, hitting the surface of the Earth can leave a large mark on Earth. Suspecting that there were major meteoroid[1] impacts in the past, planetary scientists speculated that there could be remnant marks of these impacts today. However, it was not known what shape or form these marks would have. As it turned out, the terrestrial marks left by impacts look quite similar to volcanic craters, leading to a long delay in the identification of impact craters on Earth.

One of the best known terrestrial impact craters is the crater in Arizona (Fig. 3.1). It was discovered in the late 1870s by cattlemen driving herds across the land. Its shape looked like an extinct volcano and that was what it was thought to be. The only strange thing was that this crater was ringed with pure iron. Suspecting that this may be related to a meteoroid impact, a prominent geologist, Grove K. Gilbert (1843–1918) of the US Geological Survey, visited the crater in 1892 and concluded that it was the result of a volcanic steam explosion. Gilbert built his reputation through his exploration of the American West, having traveled by foot and on mule in Utah, Arizona, and New Mexico. His studies of the geological formations of the West made him the most respected American geologist at the time, having been elected twice (1892 and 1909) as president of the Geological Society of America. Don Wilhelms, in his book "*To a Rocky Moon*", called Gilbert "one of the greatest geologists who ever lived". There are craters on the Moon and Mars named after him.

Given Gilbert's stature, his ruling on the origin of the crater was not questioned. A lone amateur, Daniel M. Barringer (1860–1929), who was a mining engineer but not a geologist, challenged this common belief and performed extensive drilling in search for an iron mass as evidence for impact. Believing that the impact must have left a large meteorite and therefore an enormous mass of iron beneath the crater, he obtained mining rights for the land in and around the crater in an attempt to mine the

---

[1] A meteoroid is a small asteroid, see definition in Chap. 2.

S. Kwok, *Stardust*, Astronomers' Universe, DOI 10.1007/978-3-642-32802-2_3,
© Springer-Verlag Berlin Heidelberg 2013

Fig. 3.1 The Barringer Crater.
The Canyon Diablo Crater, also known as the Barringer Crater, in Arizona has a diameter of 1 km and a depth of 170 m. It was created by a meteoroid impact about 50,000 years ago. Although originally believed to be a volcano, it was identified by Daniel Barringer as an impact crater. Photo credit: David Roddy and the U.S. Geological Survey

iron. Between 1903 and 1909, he drilled 28 holes to the depth of 250 m. By 1923, he has spent $600,000 of his fortune on this venture. By the time of his death in 1929, no iron mass was found. He was eventually vindicated when it was realized that the incoming meteoroid must have been completely vaporized upon impact and the crater is now named in his honor. The idea that craters are created by impact from extraterrestrial objects was not widely accepted in the geological community until the 1960s.

Since that time, over 100 impact craters larger than 100 m resulting from extraterrestrial impacts have been identified on Earth. The energy release in such impacts can rival that of a nuclear bomb. Although the bulk of the remnant of the object is melted or vaporized, fragments of it could survive and be found in the vicinity of the crater.

When we look at the Moon with a small telescope, we can see that the surface of the Moon is filled with craters. With the beginning of lunar missions in the 1960s, high quality images of the Moon obtained by lunar orbiting satellites have given us detailed morphologies and size measurements of the craters (Fig. 3.2). Why does the Moon look so different from the Earth, and where do these craters come from? For a long time scientists believed that craters on the Moon were volcanic in origin. As late as the mid-twentieth century, it was still a very popular belief in the Soviet Union that the lunar craters were created by volcanic eruptions. The first suggestion that the lunar craters are due to external impacts was made by Grove Gilbert. In October 1891, Gilbert observed the Moon for 18 nights using a 67-cm refracting telescope of the Naval Observatory in Washington.[2] He found that the largest lunar craters have sizes much larger than any crater on Earth. Lunar craters also come in many different sizes whereas volcanic craters on Earth have a definite size. This led him to suggest that the lunar craters are of impact origins. He also suggested that the

[2] The Naval Observatory is where the official residence of the Vice President is located.

Fig. 3.2 Image of the Moon. Image of the Moon taken by the *Lunar Orbiter 4* in 1967 from a lunar altitude of 3,505 km. The large crater in the middle of the picture is Schrodinger, which has a size of 312 km. Image credit: Lunar and Planetary Institute and NASA

rays that radiate from the craters are splashes from impact. His findings were published in a paper titled *"The Moon's Face"* which is now considered one of the first papers on the geology of the Moon.

By the early twentieth century, there were four competing theories on the origin of lunar craters. Suggested causes include bubbles (bursts of steam or volcanic gas), tides, volcanoes, or external impacts. In 1918, Alfred Wegener (1880–1930) performed a series of experiments on impact craters at the Physical Institute in Marburg, Germany, the results of which supported the impact hypothesis. Wegener, whose idea on continental drift was first considered crazy by the geology community but later won him worldwide recognition, also supported the impact origin of the crater in Arizona and made the bold suggestion that impact events were common in the Earth's history.

From his studies of the orbits of asteroids and comets, the Estonian astronomer, Ernst Julius Öpik (1893–1983), suggested that impacts from asteroids and comets are responsible for the craters on the Moon, and he predicted the existence of craters on Mars. The work of Öpik was picked up by the New Zealand astronomer, Algernon Charles Gifford (1861–1948), who concluded that given the high velocities of impact, the size of the impactor can be very small compared to the size of the crater. In other words, the large crater is created by the conversion of the high kinetic energy to heat and the impactor is likely to have completely vaporized in the process. Not much of the original object is expected to remain after the impact.

In 1942, the American astrophysicist, Ralph Belknap Baldwin (1912–2010), wrote two papers on the origin of the lunar craters and rays, which he attributed to the result of external impacts. His papers were rejected by astronomy journals and he had to publish them in the magazine *Popular Astronomy*. The rejection of the paper probably reflects the attitude of the astronomy community at the time, which was concerned about stars and galaxies but did not consider the Moon a worthy object of study. During the Second World War, Baldwin worked at the Applied Physics Laboratory of Johns Hopkins University on Army military ordnance. His experience with explosions and bomb and shell craters gave him the

Fig. 3.3 Crater on Mars.
Image of Mars taken by the
Mars Global Surveyor, Image
credit: Lunar and Planetary
Institute.

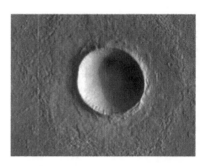

experimental verification he needed to explain the shape and depth of impact craters. In 1949, Baldwin published his work in the book *"The Face of the Moon"*. The book elaborates on the concept that the lunar craters were formed by explosions caused by external impacts.

The most important impact of Baldwin's book was its effect on the Nobel Prize winning chemist Harold Urey. This book got Urey interested in the Moon. Due to Urey's stature and influence, the Moon became the main goal of the American space program in the 1960s.

The onset of nuclear testing in the 1950s provided new opportunities for experimental studies of the impact theory. The powerful explosions of atomic and hydrogen bombs create craters which are similar to those observed on the Moon.

This debate formally ended with the Apollo landings on the Moon, where the lunar rock samples were found to contain glass formed by impacts. The existence of these impact craters suggests that the Moon was the target of large-scale bombardment by external bodies over a long period of time. Through radioactive dating of the melted rocks in the lunar samples, it was possible to establish that these impacts occurred between 3.8 and 4.1 billion years ago.

In addition to lunar samples collected by the Apollo mission, the impact record of the Moon can also be studied through the analysis of lunar meteorites. When asteroids strike the Moon, fragments of the lunar surface are ejected from the Moon. If the asteroids come in fast enough, the ejecta will escape from the Moon's surface. Some of these ejected materials end up in the Earth's orbit and eventually fall on the Earth. Since these rocks originate from the Moon, they are called lunar meteorites. It is estimated that about 0.1 % of all meteorites are from the Moon.

The use of lunar meteorites allows a widening of the geographic area beyond the small areas covered by the Apollo landings to include the polar regions as well as the far side of the Moon. All the indications are consistent with the hypothesis that the impact basins on the Moon were produced over a ~100 million year interval around 3.9 billion years ago. This event is termed lunar cataclysm by David Kring of the Lunar and Planetary Institute in Arizona. He calculated that in order to produce the basins, a total mass of about 7×10^{21} g (or 7,000 trillion tons) of external mass is needed. He believed that asteroids are the likely culprits of these impacts.

In fact, besides the Moon, the planets Mars and Mercury can also be seen to be full of craters (Figs. 3.3 and 3.4). The source of these projectiles is believed to be the

Fig. 3.4 Messenger image of
Mercury.
Many craters can be seen in
this image of Mercury taken
by the *Messenger* mission.
The brightest crater near
center of image is Kuiper.
Image credit: NASA

residual planetesimals left over from the formation of the planets. The planets are believed to have formed through the accretion process, where small pieces of rocks and ice collided and aggregated into larger bodies. After the planets were formed, the smaller kilometer-size objects, called planetesimals, were attracted by the gravity of the planets, causing frequent bombardments with the planets. Once our neighborhood was swept clean and this source was exhausted, the bombardment rate slowed down. This violent period lasted about 800 million years, and only gradually slowed down about 3.8 billion years ago. Crater records suggest that for bombardments in the first 10–100 million years of the Earth history, the intensity of bombardment is about one billion times greater than it is at present.

If the Moon was subjected to such intense external bombardments, then the Earth, being in close proximity to the Moon, could not have escaped a similar fate. Furthermore, the Earth is a larger body (therefore a larger target), as well as having a larger gravity to exert a stronger pull on the external bodies to itself. From the approach velocities of asteroids, Richard Grieve of the Geological Survey of Canada estimates that the Earth should have suffered 2.5 times the lunar impact rate per square km because of its larger surface area and stronger gravitational attraction. He estimated that there have been more than 22,000 impact craters created on Earth with diameters larger than 20 km, with ~40 impact basins 1,000 km in diameter, and several large basins with 5,000 km diameter.

Although we believe the Earth has suffered from bombardment as heavy as the Moon's, unfortunately most of the records of much of the early bombardments have been lost due to subsequent geological activities. As a planet in the Solar System, the Earth is the most geologically active one. Through the shifting of the continents (also known as plate tectonics), seas opened and closed and mountains rose up through plate collisions, and later were eroded by wind and water. These geological events have wiped out the history of external bombardment during the Earth's early years.

In spite of this, there are still over 180 impact craters on Earth left behind by collisions with large objects. The Barringer Crater in Arizona, U.S.A, with a size of over a km across, was created from impact by a 50-m size object 50,000 years ago.

Fig. 3.5 Wolfe Creek Crater.
The Wolfe Creek Crater is located on the edge of the Great Sandy Desert in Western Australia. The rim of the crater is about 25 m above the surrounding desert

When it struck, it created an explosion equivalent to about 50 megatons of TNT, or about 3,000 times the power of the Hiroshima atomic bomb. Another large crater is the Wolfe Creek Crater in Western Australia (Fig. 3.5). This 880 m, nearly circular crater was created about 300,000 years ago. Although partly filled in by sand since the time of impact, the original depth of the crater is estimated to be 150 m. A list of some of the largest impact craters on Earth is given in Table 3.1.

To discover an impact crater is not an easy task. Although there are many places in the world where the landscape may have morphological features similar to those of a crater, to prove that it is an impact crater requires evidence of impact. Usually this means the presence of minerals that can only be formed under high velocity impact conditions. Although China is a large country and there have been various suggestions of geological depressions being impact craters, there has been no definite discovery until recently. In 2009, Ming Chen of the Chinese Academy of Sciences' Guangzhou Institute of Geochemistry drilled a hole in Xiuyan in the province of Liaoning in northeastern China and found coesite, a mineral that was formed under high pressure. The shape of the quartz found also suggests that it was created by a shock event. These provided the confirmation that the bow-shape formation is indeed an impact crater, the first to be identified in China (Fig. 3.6). Calculations suggest that the size of the impactor is about 100 m and the event occurred about 50,000 years ago.

Table 3.1 Some of the largest craters on Earth

Crater name	Location	Diameter (km)
Vredefort	South Africa	300
Sudbury	Ontario, Canada	250
Chicxulub	Yucatan, Mexico	170
Manicouagan	Quebec, Canada	100
Popigai	Russia	100
Acraman	South Australia	90
Chesapeake Bay	Virginia, U.S.A.	90
Puchezh-Katunki	Russia	80
Morokweng	South Africa	70
Kara	Russia	65
Beaverhead	Montana, U.S.A.	60
Tookoonooka	Queensland, Australia	55
Charlevoix	Quebec, Canada	54
Kara-Kul	Tajikistan	52
Siljan	Sweden	52
Montagnais	Nova Scotia, Canada	45
Araguainha	Brazil	40
Mjølnir	Norway	40
Saint Martin	Manitoba, Canada	40
Woodleigh	Western Australia	40

Table adapted from Grieve, R. A. F. 1998, Extraterrestrial impacts on earth: the evidence and the consequences, Geological Society, London, Special Publications **140**, 105–131

Fig. 3.6 The Xiuyan crater in China.
The Xiuyan crater located in the Laiodong Peninsula in northeastern China was created by an impact event about 50,000 years ago. It has a diameter of about 1.8 km. The maximum height of the rim is about 200 m above the floor of the crater. In the middle of the floor of the crater is a small village. Photo credit: Ming Chen

Meteorites are the remnants of meteoroids that passed through the Earth's atmosphere and therefore are the direct evidence that such bombardments continue today. Although the rate of extraterrestrial impacts is much lower now, it is by no means insignificant. Geologists estimate that impacts producing craters of 100–200 km sizes occurred up to about 100 million years ago, although such events are now rare in modern geological times. Remnants of craters in Sudbury, Canada, and Vredefort,

South Africa suggest they had original diameters of ~250 and ~300 km, respectively, resulting from impacts that occurred about two billion years ago.

These external impacts have many implications for the historical development of the Earth. Early bombardments may have disrupted, altered, or contributed to the changing compositions of the atmosphere and the oceans (Chap. 19). Impact by an object of 0.5 km in size can create a crater of 10 km in diameter, and inject a large amount of sulfur into the atmosphere, therefore changing the atmospheric content. It has been suggested that the rich copper–nickel deposits at Sudbury, Canada could be the result of a large-scale impact that occurred 1.85 billion years ago. From model simulations, an ocean impact by a body of 0.4 km can generate tsunamis of over 60 m in height. The amount of dust thrown out by an impact that creates a crater of size 20 km is enough to block off a substantial amount of sunlight and reduce the average temperature on the surface of the Earth. Such events can happen as often as once every few million years. Larger impacts which release energy on the order of 100 million tons of TNT can cause major extinction of living species. Such impact events are estimated to occur at a frequency of once every 100 million years. There is a lot of interest in the effects of impact on the development and evolution of life, in particular whether they are responsible for the life extinction events over the history of the Earth.

The greatest prize for a crater find is to locate the site of the asteroid impact that was responsible for the demise of the dinosaurs. From fossil records, there have been at least five major extinction events in the last 500 million years of the Earth, specifically the End Ordovician event at 444 million years ago, the Late Devonian event at 360 million years ago, the End Permian event at 251 million years ago, the End Triassic event at 200 million years ago, and the End Cretaceous event at 65 million years ago. Among this "big five", the Permian/Triassic mass extinction event is considered to be the most dramatic. It is estimated that between 80 and 90 % of all the living species on Earth was wiped out. Even in the more protected marine environment, half of the species became extinct. All these happened over a relatively brief period of 10,000 to 100,000 years. Although less catastrophic than the Permian/Triassic event, the Triassic/Jurassic event removed 20 % of all marine families and many of the large amphibians. The elimination of species and the change in the ecosystem allowed the emergence of dinosaurs as a dominant group of species in the following Jurassic period. In comparison, the much more popularly known Cretaceous/Tertiary (or K/T) event that killed off the dinosaurs is relatively mild.

In 1980, American scientists Walter and Luis Alvarez of the University of California, Berkeley, shocked the world by suggesting that the K/T extinction event was the result of an extraterrestrial impact. Their theory is based on the rock sample in Gubbio, Italy, that marks the K/T boundary containing an extreme amount of iridium, which can only be brought in by an asteroid or comet of 10 km size. This theory was confirmed when similar iridium excesses were found in other K/T boundary rocks. Now the next step is to find the crater left behind by this tremendous impact 65 million years ago.

It is estimated that this crater should have a diameter of at least 200 km. The most likely candidate is the Chicxulub crater in the Yucatan, Mexico, discovered in 1990 by Alan Hildebrand (University of Arizona, now at the University of Calgary in Canada), Glen Penfield (of the Mexican oil company Pemex), and David Kring (University of Arizona, now at Lunar Planetary Institute in Arizona). This 180 km crater, buried under ~1 km of sediments under the sea is believed to be the result of an impact by a 10 km asteroid which released energy equivalent to 100 teratons (10^{14} ton) of TNT. It is estimated that the materials ejected as the result of impact had blanketed all over North America, and possibly also South America. Debris from the vaporized asteroid (or comet) spread a layer of dust all over the world. This crater was not easy to find as water and sediment on the bottom of the sea had covered most of the impact site. The finding of the crater was therefore not the result of visual observations but was based on analysis of gravity and magnetic data. The identification of the Chixulub Crater as the impact site leading to the global extinction (including that of the dinosaurs) 65 million years ago caused great excitement in the scientific community.

The exact causes of the mass extinction events are not known, with the exception of K/T, which is attributed to external impact. Other possible causes that could be responsible include global warming or cooling, nearby nova or supernova events, continental drift, massive volcanic eruptions, or ocean level changes. However, all these environmental changes could be traced back to large-body impacts, so even though the impacts may not lead to direct, immediate extinctions, the effect they have on the environment may do so.

It is now estimated that impacts leading to the creation of craters of sizes of 20 km occur at intervals of 2–3 million years, and an event similar to that of Chicxulub could occur once every 100 million years. Could the Permian-Triassic extinction event 250 million years ago be also due to an external impact? At that time, the Earth was a very different place with the continents connected in one piece of super continental land mass called Pangaea (Fig. 3.7). But over a period of less than 10,000 years, 90 % of the ocean life and 85 % of the land-based life disappeared. The possibility that this great extinction event was caused by an external impact was suggested by Luann Becker of the University of Washington. She based her evidence on the discovery of buckminsterfullerene molecules in the Permian-Triassic seabed in China and Japan. Buckminsterfullerene is a molecule made of 60 carbon atoms. It was first synthesized artificially in the laboratory, but later found to be a product of stars (Chap. 9). Furthermore, the fullerenes that were found have the isotopic signatures of non-Solar System origin. Becker suggested that "the extraterrestrial fullerenes were delivered to Earth at the Permian-Triassic boundary, possibly related to a cometary or asteroidal impact event". They estimated that the impactor has a size of 9 ± 3 km, which would be large enough to cause a massive extinction. The possibility that fullerene can serve as a tracer of extinction events is of great significance.

If the Permian-Triassic extinction event was caused by an external impact, then where is the impact site? Because of continental drift, the map of the world was very different from what it is now. At that time, the continents of Australia and

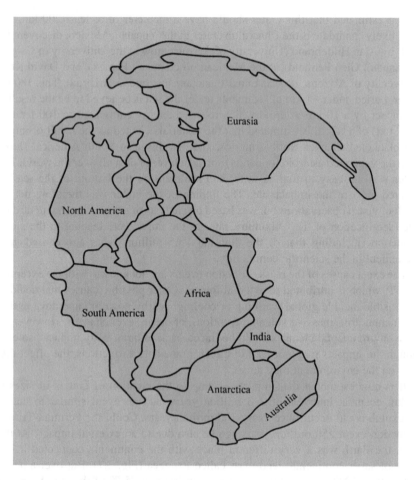

Fig. 3.7 A map of the super continent Pangaea.
All the continents used to be connected in one piece in the form of a supercontinent. The names of
the modern continents are labeled on this map. Credit: Wikimedia Commons, en:User:Kieff, File:
Pangaea continents.png, licensed under the Creative Commons Attribution-Share Alike 3.0
Unported license

Antarctica were linked together. The Bedout High, part of the Roebuck basin off the
northwestern corner of Australia, has been suggested to be the location of this
impact site.

It is also worth mentioning at this point that these extraterrestrial fullerenes were
not destroyed even in such strong impact events. This raises the possibility that
extraterrestrial organic compounds can also be carried by external bombardment to
Earth and survive. The implication of such scenarios for development on Earth will
be discussed in Chap. 21.

From geological evidence and observations of the Moon and the planets we now
know that the Earth is not a closed system. The Earth was subjected to extensive
external bombardments in the past, and continues to be bombarded, although at a

lower rate, at present. Even now, on average every year one 5–10 m meteoroid enters the Earth's atmosphere. Every one million years, we can expect a km-size object will hit the Earth. The impact events in the past are likely to have caused extensive perturbations to our physical and biological systems. To what extent the external impacts during the early history of the Earth have altered and enriched our chemical system will be explored further in the later chapters.

A brief summary of this chapter
Earth is not a closed system and was subjected to intense bombardments from the outside, bringing in extraterrestrial materials.

Key words and concepts in this chapter
- Impact craters
- Origin of craters on the Moon, volcanism or external impact?
- Impact craters on Mars and Mercury
- Impact craters on Earth
- External bombardment during the early history of the Earth
- Extinction events and their possible relationship with external impact events
- Extraterrestrial fullerenes as evidence of external impact

Questions to think about
1. How can scientists be so certain that certain meteorites originate from the Moon or Mars? What kind of evidence is needed to make this identification?
2. Mass extinction is a biological event and can be caused by a variety of reasons. Can you think of some other causes of mass extinction other than external impact?
3. If impact events capable of causing biological extinctions occur on million-year time scales, are we due for another impact? The world is much more highly populated now with human habitation covering most of the land areas of the globe. What would be the consequence if an impact does occur?
4. Geology and astronomy are considered widely different disciplines, as they study the Earth and the Heavens respectively. Can you cite some examples of how findings of one discipline influence the other?

Chapter 4
Descendants of Stars

The idea of a stellar-Earth connection has been with us for thousands of years. Many ancient cultures, both East and West, practiced astrology based on the belief that human lives are affected by the positions and movements of celestial bodies. In modern times, we have come to realize that only the nearest star, our Sun, can have direct impact on our everyday lives through its radiation and solar wind. The other stars are simply too far away to have any physical effects on us. Among our nearest celestial neighbors, only the Moon can exert enough gravitational pull to cause the tides, and the other planetary alignments, although still publicized from time to time in the popular press, do not have sufficient effect to be felt on Earth.

If our lives are not connected to the stars now, how about in the past? Is it possible that stars played a role in the beginning of life on Earth? The awareness of a physical link between stars and life on Earth was established in the 1950s when scientists learned that chemical elements have their origin in stars.

Human bodies are made up of molecules. These include simple inorganic molecules such as water, and complex organic molecules such as protein, DNA[1], sugar, and fat. Each of these molecules, no matter how large, is made up of common elements such as hydrogen (61 %), oxygen (26 %), carbon (10 %), nitrogen (2 %), calcium (0.2 %), phosphorus (0.1 %), and sulfur (0.1 %), etc. We take in atoms (in the form of molecules) by eating, drinking and breathing. We take in food, breathe in air, break them down, and use their products to fuel our activities and to build or replace our body parts. Our body is therefore a *giant chemical factory*. Chemistry, to put it simplistically, is the exchange of atoms between different molecules, with either the absorption or release of energy in the process. However, the total number of atoms remains unchanged. Atoms are not created in our bodies, nor are they destroyed.

There are 85 stable elements found in nature, ranging from the lightest element, hydrogen, to the heaviest element, uranium. Although most students who study chemistry may have trouble naming more than half of these elements, it is amazing

---

[1] DNA is the acronym of deoxyribonucleic acid, the molecular carrier of human genetic materials.

S. Kwok, *Stardust*, Astronomers' Universe, DOI 10.1007/978-3-642-32802-2_4,
© Springer-Verlag Berlin Heidelberg 2013

that many of these natural elements in fact play an important role in our bodies. If you read the label of an over-the-counter vitamin supplement bottle, you may be surprised to find that it contains minerals such calcium, chloride, chromium, copper, iodine, iron, magnesium, manganese, molybdenum, potassium, selenium, and zinc. We all know that our thyroid gland will not function without iodine. Iron is an important part of hemoglobin in our blood. Similarly, zinc is found in insulin and manganese is found in various enzymes. This chemical factory of ours in fact needs many of the natural elements to work. It is therefore clear that life relies on the availability of a rich pool of atoms to flourish.

Before we try to answer how the complex molecules in our bodies get to be what they are, we should ask where atoms come from. Are they always there somewhere in the Universe, or were they manufactured over time? If the latter is true, then when and where were they made? If they were made elsewhere, how were they brought here to Earth?

Through the use of the technique of spectroscopy, astronomers were able to determine that stars are made of the same chemical elements as the Earth. The signatures of chemical elements on Earth can be found in the spectra of stars. This shows that the heaven and Earth are made of the same materials. Even the "rare Earth elements", elements that are rare on Earth such as Scandium, Yttrium, Europium, Thulium, etc., can be found in stars.

Geologists and astronomers have different techniques to determine the relative number of elements on Earth and in stars. The number ratio of elements in stars is referred to as "cosmic abundance". In both cases, there are some obvious patterns. The most common elements are carbon, oxygen, nitrogen, neon, iron and so on. But this abundance pattern does not follow the atomic number, or the order of elements in the periodic table. While in general the heavier the element, the less abundant it is, there are clear anomalies. Iron, with an atomic number of 26, is a heavy element but quite abundant in stars and on Earth. Gold, with an atomic number of 79, is rare, but a heavier element, lead, with atomic number 82, is very common. These abundance variations suggest that there is a deeper reason behind them.

The first person to ask the question "what is the origin of the chemical elements?" was an English chemist William Prout (1785–1850). The idea behind this question is that the chemical elements did not appear as they are, but were built up over time. In terms of philosophical magnitude, this question is of the same order of "what is the origin of species", which was answered by Charles Darwin in 1859. The species did not appear as they were, but evolved from simpler forms over time. In two papers published in the Annals of Philosophy in 1815 and 1816, Prout speculated that all chemical elements were built up from the simplest element, hydrogen. Prior to Prout, people just took the existence of chemical elements for granted, or assumed that they were the creation of God, who made them available to us for our convenience. "Why are they there" is by no means an obvious question to ask. Is it possible that the chemical elements were also built up from simpler forms?

By the 1940s, nuclear physics had developed to such a state that we could realistically attempt to answer the question. George Gamow (1904–1968), a Russian

Fig. 4.1 Fred Hoyle.
Sir Fred Hoyle was a prominent astrophysicist who worked on problems of the nucleosynthesis of chemical elements and cosmology. His influence went beyond the scientific circles through his many books and radio broadcasts. Hoyle was one of those rare talents who can excel in both original research and the popularization of science. His theory on stellar nucleosynthesis is often cited as one of the greatest achievements in science in the twentieth century

born physicist working in the United States, suggested that chemical elements could be made during the early days of the birth of the Universe, during the so-called "Big Bang". However, detailed calculations showed that the conditions in the early Universe were not sufficient to produce the variety and abundance of elements as observed.

The idea of chemical elements originating from nuclear reactions in stars began with Fred Hoyle (1915–2001) in 1946. Hoyle (Fig. 4.1) was born in Yorkshire, England, and came from a modest background. His studies at Cambridge were interrupted by the War, when he worked on radar research. After the War, he returned to Cambridge and continued his academic career there, cumulating in the position as Plumian Professor of Astronomy and Experimental Philosophy in 1958. He founded the Institute of Theoretical Astronomy at Cambridge in 1967. In addition to his work on nucleosynthesis of chemical elements, he was best known for his steady-state theory on the origin of the Universe, as an alternative to the Big Bang theory. His continued political disagreements with Martin Ryle of the Cavendish Radio Astronomy group at Cambridge led to his resignation from Cambridge in 1972. After that he was basically a freelancer without a steady income. Beyond his scientific works, Hoyle wrote many books, both fiction and on the popularization of science. In his later years, his controversial ideas on extraterrestrial life made him unpopular in the mainstream astronomy circles.

No matter what others think, there is no denying his momentous contribution to the problem of the origin of elements. In the fall of 1954, William Fowler (1911–1995) from the California Institute of Technology (Caltech) in Pasadena, California, spent a sabbatical year in Cambridge, England, as a Fulbright Scholar to work with Margaret Burbidge, Geoff Burbidge and Fred Hoyle. In 1956, the

Burbidges and Hoyle went to Caltech to continue their work in nucleosynthesis. Together they worked out the nuclear reactions responsible for the creation of chemical elements, with their calculations yielding abundances comparable to those derived from astronomical observations. During the same period, A.G.W. Cameron (1925–2005) did similar calculations at the Atomic Energy of Canada in Chalk River, near Ottawa, Canada. In 1957, Burbidge, Burbidge, Fowler, and Hoyle published their work "Synthesis of Elements in Stars" in *Review of Modern Physics*. At an unusual length of 108 pages, the paper was very comprehensive. The authorship is listed in alphabetical order, which usually means equal contributions by the authors. A summary of the results was published in *Science*, with Hoyle as the lead author. Cameron's result was published in typewritten form as a report of the Chalk River Laboratory.

The fact that two completely independent groups almost simultaneously came up with the same theory on stellar nucelosynthesis is quite remarkable. While Hoyle's team worked at major research centers at Cambridge and Caltech, Cameron was very much working in isolation in a remote area of Canada. Cameron was born in Winnipeg, Canada. After he got his Ph.D. in 1952 in nuclear physics from the University of Saskatchewan, he went to work as assistant professor at Iowa State College. On the campus of Iowa State was the Ames Research Center of the United States Atomic Energy Commission. In the library of the Center, he read about the astronomical detection of the element technetium in red giant stars. Since technetium is not stable and has a short lifetime, this means that this element could not have been transported from elsewhere and must have been made right there inside the red giant star. This led to his interest in the problem of stellar nucleosynthesis and he began his calculations in Chalk River in 1954. Armed with knowledge of nuclear physics and the measured abundance of elements in stars, Cameron was able to come to conclusions very similar to those of Hoyle and company.

The calculations of these scientists showed that most of the chemical elements, with the exception of hydrogen and some of the helium, can be made in the interior of stars. The cores of stars have temperatures of tens of millions of degrees and can act as nuclear furnaces. During most of the lifetime of stars, the temperatures are such that only helium can be made. Our Sun is four and half billion years old and has been making helium all this time. As stars get older, the temperatures in the core get higher, and the synthesis of heavier elements becomes possible. The element carbon, for example, can be manufactured by nuclear fusion of helium atoms in the last moments of a star's life.

Through slow neutron capture and beta decay nuclear reactions, many heavy elements such as yttrium (Y), zirconium (Zr), barium (Ba), lanthanum (La), cerium (Ce), praseodymium (Pr), neodymium (Nd), samarium (Sm), europium (Eu), etc. are also synthesized late in the life of a star. Some elements are made under more dramatic circumstances. For example, the elements thorium (Th), uranium (U), and plutonium (Pu) are synthesized during supernova explosions.

So every atom in our body was once inside a star. It might have been created by one star, or might have been recycled through several generations of stars. By measuring the amount of each element on Earth, we found that oxygen is the most abundant element on Earth. This is, however, an anomaly. Because the gravity at the surface of the Earth is too low to keep the lighter elements, most of the hydrogen and helium has in fact escaped from Earth. When we study the Solar System as a whole, or the Sun in particular, we find that hydrogen is by far the most abundant, making up 90.1 % of all atoms in the Sun. This is followed by helium, which constitutes about 8.9 %. The next abundant is oxygen, which makes up 0.078 %. When we study the chemical composition of the Milky Way galaxy, we find that its composition is similar to that of the Sun. Since light takes time to travel, when we look at large distances, we are also looking back in time. When galaxies that are far away are examined, they are found to have less heavy elements. This confirms that the chemical composition of the Universe has changed over time, with the number of heavy elements gradually increasing as the result of stellar nuclear processing.

The understanding of the origin of the elements was one of the greatest achievements of science in the twentieth century, if not in all the history of science. In terms of importance, it ranks at a level similar to the theory of planetary motion by Isaac Newton and the theory of the origin of species by Charles Darwin. Specifically, it allows us to trace the chemical origin of all matter directly to stars, those tiny specks of light in the night sky. The atoms of carbon, oxygen, nitrogen and others are made in stars and physically spread through the Galaxy. Since our bodies are made of atoms, this puts a direct link between stars and us. Although the ancient people had imagined a star-human connection in the form of astrology, this connection through the synthesis of elements is a real and physical one. In a matter of speaking, we are all descendants of stars.

A brief summary of this chapter
Chemical elements originated in stars.

Key words and concepts in this chapter
- Our body as a chemical factory
- Stars contain the same chemical elements as Earth
- Origin of elements
- Cosmic abundance
- Stellar nucleosynthesis
- Human stellar connection

Questions to think about
1. Do you think a human body is more than just a chemical factory? If so, what do you think are the non-chemical parts?
2. Most of the physical functions of the human body can be described as results of chemical reactions. How about human behavior? Already some mental diseases can be traced to chemical imbalances. Do you think one day we can explain human behavior by physical laws?

(continued)

3. Why do we need to consider the question of the origin of chemical elements? Why can't the chemical elements just appear on Earth and in the Universe as they are?

4. The existence of chemical elements is now explained by nuclear processes. All the elements in the periodic table are synthesized from protons and neutrons. Where do protons come from?

Chapter 5
Glowing in the Dark

Before I can explain the origin of stardust and discuss what impact it might have on the rise of life on Earth, I have to go back 45 years to tell the story of the development of infrared astronomy, a new branch of astronomy that studies light invisible to the human eye. The development of instruments capable of detecting infrared light, and the use of these instruments on telescopes for astronomical observations, turned out to be the crucial element in our discovery of organic compounds in stars.

For thousands of years, our view of the Universe was limited to observations in visible light, the range of colors that human eyes can perceive. The colors violet, indigo, blue, green, yellow, orange and red, the colors of the rainbow, represent the different parts of visible light. This was first demonstrated by Isaac Newton (1643–1727), who in 1665 made a hole in a window shutter and allowed a beam of sunlight to pass through a glass prism. The light emerging from the other side of the prism had a broad range of colors like those of the rainbow. He therefore came to the conclusion that the white light of the Sun is in fact a mixture of light of many colors.

Are there other forms of light that our eyes cannot see? Actually, visible light only constitutes a small part of the total radiative output of celestial objects. Just outside of the color red lies the infrared, whose rays fill our surroundings day and night. Yet we are unaware of their presence because we cannot see them. Like blind men studying an elephant, we have been restricted to touching the trunk without access to its other body parts. Without looking at stars in colors outside of the visible range, we do not have a complete picture of the appearance and structure of stars.

The existence of energy in sunlight beyond the red was first demonstrated by the German born British astronomer, William Herschel (1738–1822). Herschel was born in Hanover but moved to England at the age of 19. He started his career as a musician and played a variety of musical instruments including the cello, the harpsichord, the oboe and the organ. He was a notable composer, having written a total of 24 symphonies as well as a number of concertos. He later developed an interest in astronomy and began observing the sky from the backyard of his house.

S. Kwok, *Stardust*, Astronomers' Universe, DOI 10.1007/978-3-642-32802-2_5,
© Springer-Verlag Berlin Heidelberg 2013

He was best known for the discovery of the planet Uranus in 1781, the first planet to be discovered beyond the five planets, Mercury, Venus, Mars, Jupiter, and Saturn, known since the ancient times.

In 1800, Herschel split sunlight into a color spectrum by passing the light through a prism and placed three thermometers under different colors of the split sunlight and measured their temperature. He noted that the highest temperature was recorded under red light. However, when he placed a thermometer beyond red light, he found that the temperature of that thermometer was higher than the temperatures of thermometers not placed under sunlight. He therefore detected energy in light beyond the visible.

Whether the experiment performed by Herschel led to the understanding that sunlight consists of redder colors than we can see with our eyes was a different issue. At that time, it was a common belief that light and heat are two different things and Herschel thought that his experiment had confirmed the separation of these two entities. While he discovered heat (called caloric rays at that time) in sunlight beyond the red color, he should not be credited with the discovery of infrared light.

Heat and light

As our eyes can respond to light and our ears can hear sound, our body can also sense heat. Our everyday feelings of "hot" or "cold" are responses to the random motions of molecules in the air. The faster the molecules are moving, and therefore the more frequently they hit our body, the hotter we feel. Since this form of heat is associated with motion, the temperature associated with it is called "kinetic temperature".

However, we can also feel heat, even in the absence of air. We feel the heat of the Sun even though the Sun is separated from the Earth by huge distances over a near vacuum. At high mountain tops where the air is half of what it is at sea level, the warmth of the Sun is the same, if not more. Scientists in the eighteenth and nineteenth century recognized this other form of heat, and it is referred to as "heat rays" to distinguish it from "light rays" which can also be transmitted over vacuum. To emphasize the difference between these two kinds of radiation, scientists at that time noted that while the Moon is also bright, it does not carry heat and therefore moonlight was considered "cold light" while the sunlight is "warm".

The next celestial object detected in the infrared was the Moon. In 1846, Macedonio Melloni (1798–1854) used a 1-m lens to observe the Moon from his Naples balcony and detected heat from the Moon. It was well known that visible light from the Moon is the result of reflected sunlight, but careful measurements with a telescope showed that the Moon radiates energy in the infrared which is quite different from reflected light from the Sun. In fact, this infrared radiation is not reflected sunlight but light emitted by the Moon itself.

The unification of the concepts of light and heat did not occur until the nineteenth century. We now realize that there are different kinds of light (or more precisely electromagnetic radiation), and they can be in the form of X-ray, ultraviolet, visible, infrared, microwave, or radio. What distinguish them are their wavelengths (or colors for visible light). Heat is a consequence of interaction between light and matter. When sunlight strikes our skin, we have a sensation of warmth. This sense of heat is not the result of a separate caloric ray from the Sun. This was summarized in the book *Light, visible and invisible* by Silvanus P. Thompson (1851–1916) in this way: "The chief physical effect produced by these long infrared waves is that of warming the things upon which they fall". However, a full understanding of the relationship between light and heat only became possible after the formulation of quantum physics, and a new understanding of the quantum interactions between light and matter.

Setting aside the theoretical understanding of the nature of infrared light, technology for the detection of infrared light continued to improve in the twentieth century. In order to measure the brightness and distribution of infrared rays, we have to rely on instruments that can respond to and detect them. Part of the motivation for the development of infrared detectors originated from the military, where the ability to observe enemy soldiers in the darkness provides a large tactical advantage. The U.S. Air Force was also keen to be able to use infrared signals to distinguish incoming enemy intercontinental ballistic missiles from normal background stars. Astronomers were quick to take advantage of this new technology. In the late 1960s, astronomers mounted infrared detectors on telescopes and tried to search the sky for celestial sources of infrared radiation. Our newly acquired capability to observe the Universe in the infrared led to many surprises, not the least of which is how stars could have influenced the development of life on Earth.

Heat Ray and Chemical Ray

Herschel believed that the thermometer measurement beyond the red was the result of a separate "heat ray". This issue of different kinds of sunlight was further complicated by the discovery in 1801 by Johann Wihelm Ritter of another radiation beyond the violet end of the solar spectrum that causes chemical changes. This was given the name "chemical ray". The belief that the Sun emits "chemical", visible, and "heat" rays persisted throughout the nineteenth century.

The difficulty arose because of the different effects of light on us. Light is reflected off a surface and therefore we notice the effect of "illumination" of light with our eyes. We feel the "heat" through our sense of touch, and we can detect chemical changes induced by light. By noting the different effects of light, scientists erroneously concluded that they have different origins. In fact, matter responds to light through both reflection and absorption, and

(continued)

responds differently to different colors of light as the result of the quantum structure of matter.

This confusion was not sorted out until many experiments later demonstrated that the solar "heat ray" (or infrared light) has the same polarization, interference, and absorption bands as visible sunlight. The concept of "heat ray" therefore dropped out of scientific usage into the domain of science fiction, as used by H.G. Wells in "The War of the Worlds".

Vision is the most powerful of our five senses of hearing, touch, smell, taste, and sight. Most of the information we gather in our everyday life is derived visually. We can discern objects because objects reflect sunlight onto our eyes, which generate a signal to our brain. Sunlight is made up of light of all colors, from red to violet. This is illustrated by rainbows where water vapor in the air disperses sunlight into its color components. But physical objects do not respond equally to different colors of light. They absorb some colors and reflect others. Leaves are green because they reflect most of the green light but absorb the red and the blue. Charcoal is black because it absorbs all colors well and reflects little, whereas snow is white because it practically reflects all the incoming light. In winter, we prefer a dark-colored coat because of the heat generated by the absorbed sunlight, whereas in the summer white clothing is preferable because it deflects all the energy of sunlight.

One of the most fundamental principles of physics is that energy is conserved. When light is absorbed, its energy must go somewhere. Most of the time, the light energy is converted to heat. As a result of sunlight shining on a pond, the water molecules become agitated and move faster. Similarly, light shining into a room through a window causes the air molecules to move faster in random directions. When these fast moving air molecules hit our skin, we get a sensation of heat. In order to quantify this sensation, physicists make up a concept called temperature. A sensation of warmth corresponds to high temperature which in turn corresponds to a collection of molecules moving fast.

The commonly used temperature scales of Fahrenheit and Celsius (or centigrade) are artificial scales created by people to quantify our sense of warm and cold. Fahrenheit is a temperature scale made up by Daniel Gabriel Fahrenheit (1686–1736) of Holland in 1717. The temperature scale nicely covers the range of temperatures encountered in Europe. One hundred degrees would correspond to the peak temperature in the summer, whereas zero would approximately correspond to the coldest day in the winter. In 1742, a Swede called Anders Celsius (1701–1744) devised a more "scientific" scale that makes use of the boiling point and freezing point of water. The former is defined as 100° in temperature and the latter as zero. The Celsius scale is not intrinsically better or worse than the Fahrenheit one; the two scales were just designed for different purposes. For better or for worse, the Fahrenheit scale, which is more convenient for ordinary people, has now largely been abandoned in almost all countries except the United States.

Later when it was realized that temperature is in fact related to the random motion of molecules in the air, physicists knew that there must be a temperature at which all molecules stop moving. This is referred to as "absolute zero". The Kelvin temperature scale (named after Lord Kelvin, 1824–1907) uses absolute zero as the starting point, and retains the intervals of the Celsius scale. Therefore zero degree Celsius becomes 273° K, and the boiling point of water becomes 373° K. Since the Kelvin scale is not arbitrary and has a physical basis, it is considered a more rational temperature scale and is commonly used in physics, chemistry and astronomy.

Although solid substances do not contain fast-moving molecules as those in air, they are also made of atoms. Some simple solids are made of atoms arranged in a highly organized regular form. For example, salt is structured in a simple cubic form with the sodium and chlorine atoms located at alternate corners of the cube. Other solids, e.g. rock, may have more complicated or even random structures. No matter what the structure is, when light enters a solid, it causes the atoms to move and warm up.

There is another principle of physics: "anything that has a finite temperature above absolute zero will shine". We now know that ultraviolet, visible, and infrared light are all part of the electromagnetic radiation emitted by the Sun. In fact, like any warm body, the Sun radiates throughout the entire electromagnetic spectrum from the X-ray to the radio, and the fraction of light that is emitted in each region is only dependent on the body's temperature. The hotter the object, the more its radiation is shifted to the blue, and a cooler object will radiate more in the red. The Sun has a temperature of about 6,000° K, and most of its light is radiated in the visible region. Our bodies are much cooler, about 37° C (or about 310° K), and we shine mostly in the infrared (Appendix C). Not only living things like animals and trees shine, but also rocks and stones, tables and chairs. This may seem surprising. Our visual perception is based mostly on reflected light. When we move a chair to a windowless dark room, we can no longer see the chair. This does not mean that the chair does not give off light; it is just giving off light in a color that we cannot see!

For everyday objects, they are luminous in two different ways. They are bright in the visible because of reflected sunlight and at the same time they are also self-luminous as they are also radiators of infrared light. While the Sun shines in all colors, the Moon is bright in the visible to varying degrees at different phases because its visual brightness is due to reflected sunlight. Only the parts of the Moon facing the Sun will be visually bright, but the whole Moon is always bright in the infrared, even during the New Moon.

The response of our eye vision is limited to colors between red and violet. For light that is bluer than blue (called ultraviolet or UV light) or redder than red (called infrared light), we simply cannot see it. There is no intrinsic difference between red and infrared light (or between blue and ultraviolet light); the only difference lies in the limited response ability of our eyes. The kind of light that is given off by warm objects of room temperature (about 27° C or 300° K) is mostly infrared light. Since there are many warm objects all around us, we are surrounded by infrared light night and day. If our eyes responded to infrared light rather than visible light, there would be no difference between night and day because the Earth would be warm and bright in the infrared at all times. We could still see the Sun and the Moon, but they would no longer be the dominating light sources. Objects instead of reflecting sunlight would in fact glow by themselves. We would all look like walking ghosts!

Another consequence of infrared vision is that the science of astronomy would probably be delayed for thousands of years. Due to the diminished difference between night and day, we would no longer have the advantage of darkness to view the stars. Overwhelmed by infrared light from the ground, we would have a hard time making out the faint stars. Instead of astronomy being the earliest discipline of science developed, humans would have been unaware of the Universe until sufficiently sensitive light detecting technology at other wavelengths was developed.

This is entirely hypothetical, of course. Humans are not meant to have infrared vision. The Sun, being a star of about 6,000° K, radiates primarily visual light. Most animals took advantage of this fact and, through the process of evolution, adapted their eyesight to respond to sunlight. Certain snakes, however, do specialize by having infrared eye sight to catch their prey at night. One could speculate if the Earth had been hit by a giant asteroid or a comet which threw out a huge amount of dust that covered the Earth for millions of years, the world would have been thrown into perpetual darkness. Most animals would have lost their visual advantage to see and avoid predators. Would snakes have gained enough advantage in that constant darkness to become the dominant species on Earth?

Most of the stars in our Milky Way galaxy are like the Sun. Yellow stars are very common in our Galaxy. Some are redder and some are bluer, but very few are so red or so blue that we cannot see them. For this reason, for thousands of years, our perception of the Universe has been dominated by stars. If infrared astronomy, not visible astronomy, were the first to be developed, how would our view of the Universe have changed?

We would still see the Sun, the Moon, and the planets. Being near us, they outshine everything else in the sky in the infrared. There would still be stars in the sky, but in addition to these points of light, there would be extended patches of light. The Milky Way would be very prominent, not because of the stars but because of clouds of dust. Much of the light in the sky is due to radiation from dust—dust in isolated clouds or dust surrounding stars. Because our Sun is located near the edge of the Galaxy, there is a concentration of clouds in the direction of the Galactic Center in the constellation of Sagittarius. The brightest point-like sources that appear like stars are in fact not star light in the strict sense of the word, but light given off by dust in the surroundings of stars.

The first survey of the infrared sky was done at the California Institute of Technology in the 1960s. The technology available at the time limited the survey to infrared colors not much redder than the red end of visible light, which we now refer to as near-infrared light. Using a 62 in. (1.6 m) telescope made of aluminum and an infrared detector first developed for military use for heat-seeking missiles, Gerry Neugebauer and Robert Leighton surveyed the northern sky at the wavelength of 2.2 μm and catalogued 5,612 infrared sources. Since normal stars like the Sun also radiate in the near infrared, skeptics at the time predicted that nothing interesting would emerge from the survey except a few red stars.

Although the Neugebauer–Leighton infrared catalogue (or IRC for short) does contain many well-known stars such as Vega, Altair, and Betelgeuse, it also revealed a number of previously unknown new stars. Some of these new infrared stars are so red that they can hardly be seen in visible light. The existence of stars of such red color is totally contradictory to our theoretical understanding of the structure of stars. The coolest normal stars have temperatures of about 3,000° K, but some of the stars found by Neugebauer and Leighton have temperatures as low as a few hundred degrees!

The most extraordinary example is the object numbered +10216 in the IRC catalogue. It is one of the brightest infrared sources in the sky, and yet it is so faint in visible light that it can only be seen with a large optical telescope. Since it is located in the constellation Leo (the lion) and its light output is variable, it has since been given the official variable star name CW Leo. It appears to have a temperature of only 540° K, much cooler than any stars known at that time.

It is interesting to note that the star CW Leo, the brightest star in the sky at the wavelength of 5 μm, is extremely faint in visible wavelengths. The reason for its optical faintness is that most of the visible light that the star emits is obscured by the dust particles the star itself ejected. The star is ejecting so many dust particles that these particles are able to absorb all the star light. As a result of absorbing the energy of the star light, the dust particles are heated up and begin to radiate in the infrared. So while the star is visibly faint, it is extremely bright in the infrared. Without having done a survey of the sky in the infrared, we would never have been aware of the existence of these stars.

The Neugebauer–Leighton survey demonstrated that there is much to be learned by observing the infrared sky. By the late 1960s, detectors were developed to pick up radiation much redder than those used by Neugebauer and Leighton. However, at these colors (which we call the mid-infrared, light in the wavelength range of 10–25 μm), the Earth's atmosphere is an increasing problem. Water vapor in the atmosphere blocks off much of the infrared light from space. Fortunately, there are a number of windows, e.g., around the wavelengths of 10 and 20 μm, where the atmosphere is less opaque. It is possible to carry out ground-based observations, in particular at high mountaintops, through these atmospheric windows.

In order to gain a better understanding of the infrared sky, the U.S. Air Force Geophysical Laboratory (AFGL) put telescopes equipped with mid-infrared detectors on rockets. Since rockets can rise high above most of the Earth's atmosphere, these telescopes can have a much better view of the infrared sky. The use of rockets for astronomical observations began soon after the U.S. captured a number of V-2 rockets from Germany after World War II. Since ultraviolet and X-ray light cannot penetrate the Earth's atmosphere, rockets were used by the U.S. Naval Research Laboratory to observe the Sun in ultraviolet and X-ray in 1946 using

Fig. 5.1 The Aerobee rocket
used in the AFGL sky survey.
The Aerobee 170 rocket has a
length of 31 ft and weight of
1,850 lb. It can carry a 250 lb
payload to an altitude of 165
miles. Photo credit: White
Sands Missile Range Museum

the V-2 rockets. This tradition of astronomical observations using rockets was extended to the infrared by both the U.S. Navy and Air Force (Fig. 5.1).

Between 1971 and 1974, the AFGL team launched a total of seven rocket flights from White Sands Missile Range to survey the northern sky and two from Woomera, Australia, to cover the southern sky. The Aerobee 170 rockets took the telescope to an altitude of ~160 km in the northern program and the Aerobee 200 rockets were used in the south, taking the telescopes to an altitude of ~190 km. On each flight, about 200 s worth of data were taken in the northern program and ~275 s in the southern program.

Together, these rocket flights managed to observe 90 % of the sky in the mid-infrared. This first mid-infrared survey produced a catalogue of 3,198 sources published by Russ Walker and Stephan Price of AFGL in 1975. Many of these sources have no known visible counterparts. The task of finding the counterparts of these new sources was contracted to three universities: the University of Minnesota, the University of Wyoming, and the University of California at San Diego. Since the rocket observations had limited pointing accuracy, the positions of the infrared sources were not well determined. Ground-based telescopes, on the other hand, could point accurately, and by searching in the general directions indicated by the rocket observations, the exact positions of the infrared sources could be determined. In this way, one could also see if there was an optical counterpart at that position. Sometimes, the counterpart was not a point-like star, but a nebula with extended structures. Examples of new nebulae discovered by the ground-based identification process include AFGL 618 and AFGL 2688, both of which turned out to be proto-planetary nebulae (Chap. 8). The object AFGL 915, later named the "Red Rectangle", is another exotic object (Chap. 13).

For star-like objects, there was AFGL 3068, which was found to be similar to the carbon star CW Leo, except that it is even cooler. The best estimated temperature of this object is 320° K. The discovery of strange infrared stars such as CW Leo and AFGL 3068 clearly suggested that we did not have a complete understanding of stars. Since the theory of the structure and evolution of stars has been worked out in the 1940s and 1950s, astronomers have gotten into a state of comfortable complacency. These stars served as a wake-up call that more work was needed.

In the chapters that follow, we will find that these peculiar stars provide two important clues to our story. One is that their very low temperature is the result of the stars being surrounded by shed dust, showing that the stars are manufacturing stardust at a very high rate. The second is that these are among the most carbon-rich stars in the Galaxy. The fact that their internal nuclear reactors are producing a large abundance of carbon provides for a favorable environment for the synthesis of organic compounds.

A brief summary of this chapter
Infrared astronomy allowed scientists to see light invisible to human eyes and discover new kinds of stars.

Key words and concepts in this chapter
- Discovery of infrared light
- Reflection, absorption and emission of light by physical objects
- Light as a form of electromagnetic radiation
- Every object with a finite temperature will radiate
- Relationship between color and temperature
- Spectroscopic signatures of matter
- Development of infrared detectors
- Infrared sky surveys
- Infrared stars

Questions to think about
1. Our eyes can detect light and our bodies can sense heat. Why did we develop these capabilities? What are the evolutionary advantages that led to our possession of these capabilities?
2. It is hot near the tropics and cold near the polar regions on Earth. What are the physical processes that lead to these temperature variations? Even at the same latitude, it is colder on a high mountain than at the foot of the mountain. Why is this?
3. If we had X-ray vision rather than visual vision, what would our world be like? Would it be an advantage to have X-ray vision?
4. The fact that we can see colors means that our eyes are in fact crude spectrometers. If our eyes were only capable of differentiating different brightness but not color, how would our lives change?
5. Is it a coincidence that stars are bright objects in visible light?
6. How would the development of astronomy have changed if we had infrared vision but could not see visible light? Consider this question beginning with early human civilization without technology and then to different degrees of advancing technology.

(continued)

7. Almost every country in the world has adopted the Centigrade scale for the measurement of temperature, with the exception of the United States which uses the Fahrenheit scale. Which system do you think is better? If you are an American, do you think the United States should switch to the Centigrade scale?

8. The US Navy and Air Force both played major roles in the development of astronomy. What do you think of the role of military research in science?

Chapter 6
Stardust in Our Eyes

Since infrared detectors became new toys for astronomers, they have been used to observe a variety of celestial objects. In 1965 there was a bright comet called Comet Ikeya-Seki. In the fall of 1965, it became visible in the dawn sky of the northern hemisphere and Eric Becklin and Jim Westphal of the California Institute of Technology used the 24-in. (60 cm) telescope on Mount Wilson to observe this comet in the infrared. Like planets and asteroids, comets derive their brightness from reflected sunlight. Therefore the closer a comet gets to the Sun, the brighter it becomes. The color of comets therefore also resembles the color of the Sun.

While this is true from blue to red to the near infrared, when Comet Ikeya-Seki was observed in the mid-infrared, it was found to be much, much brighter than reflected sunlight could account for. The comet was radiating on its own in the mid-infrared! Now recall our previous discussion on the solid objects in our everyday life. They absorb sunlight and re-emit this energy in the infrared. The same must be true for comets. Since the 1950s, it has been popularly assumed that comets are nothing more than "dirty snowballs". However, the infrared observations suggested something more. In addition to the gaseous molecules that make up the comet, there must also be solids inside.

The existence of solids in interstellar space has been known since the early twentieth century. When pictures are taken of the Milky Way, certain dark patches can be seen among the dense field of stars (Fig. 6.1). At first, these dark patches were referred to as "holes" or "voids" because they seemed to suggest an absence of stars. William Herschel referred to these dark patches as "openings in the heavens". The American astronomer Edward Emerson Barnard (1857–1923), who catalogued many of these objects in his *Atlas of Selected Regions of the Milky Way*, referred to them as "dark markings in the sky". He also called them holes in the sky and he even coined the term "black holes" to refer to these dark patches, although the term "black hole" has very different meaning nowadays. The German astronomer Max Wolf (1863–1932) called them "dark caves" and "black voids". It was not until the 1930s that astronomers figured out that these dark patches in the sky are not due to the absence of stars, but the result of solid particles in interstellar space obscuring the starlight behind. These particles were given the name "interstellar dust".

S. Kwok, *Stardust*, Astronomers' Universe, DOI 10.1007/978-3-642-32802-2_6,
© Springer-Verlag Berlin Heidelberg 2013

Fig. 6.1 Dark cloud Barnard 68. The dark patch in the middle of this image indicates a concentration of interstellar dust. The absence of stars is because light from the background stars is obscured by the foreground dust cloud. The presence of such dark clouds in the Milky Way allowed us to learn about the existence of interstellar solid particles. Image credit: European Southern Observatory

Solid particles of different sizes absorb visible light of different colors differently, with blue light being affected more than red light. If a star suffers from obscuration by interstellar dust between us and the star, the star appears redder than it should. Through a survey of stellar colors, one can discern how much dust is present along the line of sight to the stars. By the mid-twentieth century, astronomers came to accept that solid particles are widely present in interstellar space in our Galaxy.

Astronomers generally regarded interstellar dust as a nuisance as it dims the visible light of distant stars and affects our ability to measure the brightness of stars accurately. The serious study of dust itself had to wait for the development of infrared astronomy, when the self-radiation of these particles could be detected directly (Fig. 6.2).

How big are these solids in comets? Are they of the size of rocks, sand, or baseballs? By observing comets at different infrared colors, astronomers were able to derive the temperature of Comet Ikeya-Seki to be about 1,150° K. For comparison, the melting point of iron is 1,800° K. While this temperature is much lower than the temperature of the Sun (about 6,000° K), it is nevertheless much warmer than the Earth (about 300° K). Since comets, planets and asteroids are all heated by sunlight, their temperatures depend on how far away they are from the Sun. This is something we have also experienced: the closer you are to a fireplace or a campfire, the warmer you are. Since the comet's distance to the Sun was known, astronomers could work out from the principles of physics how warm the comet ought to be. This temperature turns out to be 750° K, or 400° cooler than the 1,150° temperature as the color of the comet suggests.

An object settles at a certain temperature as a result of the rate that it is heated and the rate that it cools itself. This high temperature of the comet means that the

Ed Ney began his career in nuclear physics separating uranium 235 and later moved onto measuring cosmic rays in balloon experiments, then to the study of atmospheric and solar physics from balloons. His expertise in balloon flights put him into contact with Martin Schwartzschild of Princeton University, who in the early 1960s was in charge of Project Stratoscope II. This Princeton connection got Ney interested in infrared astronomy, which at that time was a completely unexplored subject. Ney picked up the techniques of infrared instrumentation from Frank Low of the University of Arizona and then proceeded to start an infrared astronomy group at the University of Minnesota.

Ney was a colorful character. He would jump on top of tables when he lectured classes and often cursed and swore while he observed on the telescope. Once, on the way back from the *O'Brien Observatory,* he drove his powder-blue Jaguar XKE at twice the 65 miles per hour speed limit to see how large a fine he would get. To everyone's disappointment, the Minnesota State Highway Patrol failed to appear. Woolf is an Englishman who has his own unique qualities. It was rumored that he was so focused on his work that he left many of his paychecks in his office drawer uncashed.

With this infrared telescope, they began to observe stars and comets. It was there that many interesting discoveries were made. They got one of their first surprises observing red giants. Red giants are stars that are ten to a hundred times larger and hundreds to thousands of times brighter than the Sun. When our Sun gets old, it will become a red giant (Chap. 7). The older a red giant is, the redder and more luminous it becomes. When Woolf and Ney observed some very old red giants, they found that they had more infrared light than expected. Drawing on the example of comets, they concluded that there must be small solid particles in the atmospheres of these red stars. This was the first concrete identification of stardust.

By making more precise color measurements, they found that the infrared emissions from these stars had a very strong feature at the wavelength of 10 μm. This feature is characteristic of silicate rock on Earth. By comparing the star's spectra to the infrared spectrum of terrestrial rock samples compiled by John Gaustad of the University of Colorado, the Minnesota astronomers concluded that the stardust in red giants must also be made of the same stuff!

This was in fact quite a bold suggestion and was viewed with skepticism by other astronomers. Everyone knew that stars are made of gases and the temperatures at the surface of stars are thousands of degrees. The red giants, although cooler than most stars, still have temperatures of about 3,000° K. Surely a rock would melt at such a high temperature. To make a piece of rock, even a very small one, requires putting together thousands of silicon and oxygen atoms (the basic ingredients of silicates) with other metals such as magnesium and iron. Given the fact that the density of stellar atmospheres is billions of times lower than that of the terrestrial atmosphere, bringing together thousands of atoms in a single place to make a piece of solid seemed an impossible task.

Again, comets came to the rescue. In 1969 came Comet Bennett. It was observed with the Minnesota telescope and found to have the same silicate feature as in red giants! This was clearly no fluke. Comets, like the Earth, are objects of the Solar System. Silicate rocks make up most of the Earth's mantle, and probably much of the Moon's. In the Earth's crust, 47 % of all the atoms are oxygen, followed by silicon, which makes up 28 %. It is therefore much easier to accept the presence of silicate dust in comets. Since comets and red giant stars have the same infrared signature, silicate dust must also be present in the atmospheres of red giants.

Further observations were made by Bob Gehrz, then a graduate student at the University of Minnesota. He found that the silicate feature is present in many old red giants. When stars get very old, they become unstable and begin to pulsate, with their radii expanding and shrinking periodically. This class of pulsating stars is named after its most famous member, Mira, and is known as Mira Variables. Gehrz found that almost all Mira Variables show this silicate feature, suggesting that stars begin to manufacture silicate dust when they get old. For the first time, we had solid evidence that solids are being made by stars.

With these successes, the University of Minnesota collaborated with the University of California at San Diego to build a larger 60-in. (1.5 m) telescope on Mt. Lemmon near Tucson, Arizona. The Mt. Lemmon site was first developed by the U.S. Air Force as a radar station as part of the Air Defense Command. At an altitude of 2,800 m in dry Arizona, Mt. Lemmon served well as an infrared observing site. Observing with telescopes such as the *O'Brien* and *Mt. Lemmon observatories* was quite an experience. One had to stand all night on a ladder inside a freezing-cold dome to manually guide the telescope. Since the pointing of the telescope tended to drift as the telescope tries to follow the moving stars through the night, the observer had to constantly view a nearby bright guide star to keep the telescope pointing in the right direction. This was done through a hand-held controller, similar to controllers used in video games today. The poor observer had to constantly bend over to look through the finder telescope,[1] fiddle with the guiding controls, and at the same time try to stay awake and not fall off the ladder. The dome and the telescope were operated by hand. When the telescope or the instrument broke down, it had to be fixed by the observers as there were no observing assistants. The demanding conditions of infrared observing were another source of frustration as we were at the mercy of weather. It is a great contrast to the modern comfortable way of observing sitting in a warm, heated control room while allowing the computer to take charge of many tasks which previously had to be done by hand.

As incredible as the discovery of stardust was, it is even more amazing that this was foretold by Fred Hoyle in 1955, 14 years before the discovery. In his book, *Frontiers of Astronomy*, Hoyle wrote

---

[1] A finder telescope is a small telescope attached to the main one to allow the observer to locate the star field. A finder telescope has a larger field of view. The observer compares what he sees through the finder telescope to the finding chart (a printed star chart prepared previously) that he holds in his hands in order to move the telescope to the correct direction towards the intended target.

"Dust particles originate in the atmosphere of stars of low surface temperature. It can be shown that at temperatures below about 2,000°, carbon atoms in the atmosphere of a star will not remain gaseous but will condense into solid particles. It can also be shown that when the particles grow to about the wavelength of blue light—about one-hundred-thousandth of an inch—the radiation from the star pushes them outwards even in spite of the inward gravitational pull of the star" (Hoyle 1955, p. 240).

Hoyle, having worked out the theory of stellar nucleosynthesis (Chap. 4), realized that the element carbon is made by burning helium in the late stages of the life of stars. The carbon atoms, if brought to the surface by convection from the core of the star, may condense into solids. This was quite a bold suggestion as it is well known that red giant stars, being so big, have very dilute, low density atmospheres. By terrestrial standards, the density is so low that it qualifies as a vacuum in laboratory conditions. Since solids form when atoms stick to each other, it takes a very long time before micrometer-size solid particles can form. This growth process is also competing with the process of evaporation. If the temperature is high, solids will melt or evaporate, therefore preventing them from growing to an appreciable size. The kind of grains that Hoyle envisioned was carbon grains. If such carbon grains are like graphite, which is a metal, they absorb star light very efficiently and therefore heat up very quickly. This is not conducive to their creation in stellar atmospheres.

While scientists still debate how solid grains form in the atmospheres of red giant stars, there is no doubt from observations that they are being made in large quantities over the very short time scale of tens of years. We now know that the grains are not formed in the atmospheres, but above them. At higher altitudes, the temperature is lower and the conditions more favorable. Under low density conditions, the temperature we feel is not so much the temperature of the air, but the temperature of the radiation. One can experience this on high mountain tops. Even if the temperature is as cold as $-30°$ C, a mountain climber can still feel warm because of radiation from the Sun. In a stellar environment, Robert Gilman of the University of Minnesota was able to show that the temperatures of the solid grains are mainly determined by how far away it is from the star. Also he noted that materials like glass that are transparent in the visible but radiate strongly in the infrared can cool more quickly than metals. Silicates, being similar to glass, in fact can reach condensation temperatures just a couple of stellar radii above the stellar atmosphere.

The different responses to visible and infrared light by glass can be illustrated by the familiar green house. Since glass is transparent to visible light, a green house is able to let in the sunlight. However, glass is opaque to infrared light, so when the plants inside the green house re-emit the sunlight that they absorb in the infrared, the infrared light is trapped inside the green house, keeping it warm. The silicate grains in a stellar atmosphere work like an inverse green house. Being near transparent to visible light, the grains are poor in absorbing the star light. However, they are good in absorbing (and therefore emitting) infrared light, and therefore can radiate away their energy much more efficiently. By being able to keep cool, solids can condense.

This principle illustrates that if any solid is expected to form in the stellar environment, silicate is a natural candidate. However, this does not explain how thousands of atoms can get together in one place and cluster themselves into a solid. Since the densities in the atmospheres of red giants are lower than the best vacuum that scientists can create in the laboratory, atoms collide with each other only rarely. How they can end up as solids is still a mystery.

Although we do not understand theoretically how stars manage to make silicate dust, we do know from observations that old stars are able to do it efficiently and are constantly ejecting these dust particles into space. Furthermore, silicates are not the only type of dust made by old stars. There are two kinds of old red giants; the first kind has more oxygen than carbon (called oxygen rich), and Mira Variables are examples of this kind. The others are those that have more carbon than oxygen (called carbon rich), and they are usually called carbon stars. The silicate dust is found to be present exclusively in oxygen-rich stars. One example of oxygen-rich red giants is Betelgeuse, the brightest star in the constellation of Orion. In 1974, Richard Treffers and Martin Cohen of the University of California at Berkeley, observing with the 2.2-m University of Hawaii telescope on Mauna Kea, detected an infrared feature at the wavelength of 11.3 μm in the spectrum of carbon stars. This is clearly different from the silicate dust. By comparing these results with measurements in the laboratory, Treffers and Cohen concluded that this feature originates from a solid compound called silicon carbide (SiC). Silicates and silicon carbides turned out to be the first of many minerals to be manufactured by stars and the discovery of these two substances represented the beginning of a new field we now called astromineralogy. Much of the experimental work on silicate stardust was done by Thomas Henning's group at Jena University in Germany and the group at the Institute of Astronomy of the University of Vienna in Austria. These experimental studies have turned out to be extremely valuable in the identification of minerals in the spectra of stars.

The beginning of infrared astronomy in the late 1960s not only saw the discovery of stardust, small solid particles made by stars, the era also marked the beginning of identification of what kinds of solids they are. This was accomplished by comparing the astronomical infrared spectra with laboratory measurements of known minerals on Earth. These results show that stars are not just made of gases, but can also manufacture solids as well. The similarity of stardust and the rocks and minerals on Earth shows that common chemical processes are at work and stars and the Earth have more things in common than we previously thought.

In the remaining chapters of this book, we will discuss how astronomers are continuously surprised by the degree of complexity of these tiny solids made by stars. Not only do they contain minerals, they also contain diamonds and other rare forms of gems (Chaps. 11 and 12). Most unexpected of all is the discovery of complex organics found in stardust. Some stars are able to make and spill out organic solids with structures that rival in complexity those of coal and oil (Chaps. 10 and 14). There is also the possibility that these stellar solids may have a chemical composition which is unknown on Earth, and some unexplained light phenomena in the Universe (Chap. 13) could be attributed to these new forms of solids. These discoveries were all made possible by the development of infrared astronomy.

A brief summary of this chapter
Stars make dust (small solid particles) that is similar to rocks on Earth.

Key words and concepts in this chapter
- Discovery of stardust by infrared astronomy
- Solid particles in interstellar space
- Size of cometary dust particles
- Identification of nature of stardust by infrared spectroscopy
- The formation of silicates and silicon carbide solid particles in old stars
- Similarity between stardust and rocks on Earth
- Astromineralogy

Questions to think about
1. Is it surprising that minerals are also found in stars? Why did astronomers find it hard to accept the existence of minerals in stars?
2. Our household dust contains a wide mixture of things, many of which are products of life. Is it possible that stardust also contains products of life?
3. We can learn a lot about the composition of household dust by visually examining it under a microscope and by performing all kinds of physical and chemical tests. But we cannot do the same for stardust. How can we know what stardust is made of?
4. The ancient people believed that stars are very different from the Earth. In what ways do you think stars and Earth are similar and different?
5. The Sun provides an example of what stars look like. Recent space missions have provided us with up-close images of the Sun. The Sun is found to have dark spots, flares, lobes, prominences, corona, magnetic fields, storms, and winds. If the Sun is a typical star, can we claim we understand stars?

Chapter 7
The Oldest and Brightest

Stars, like human beings, do not live forever. They get old and they die. For most of a star's life, e.g. where the Sun is now, it does not change much. We owe our existence to the fact that the Sun has been pretty much stable for four and half billion years, allowing a steady environment for life to have developed and flourished on Earth. Had the Sun's brightness or temperature been changing, even on million-year time scales, the development and sustaining of life would be much more difficult. However, when they get old, stars become redder, brighter, and larger. In about five billion years, the Sun will expand to the orbit of Venus and will be several hundred times brighter than it is now. The Sun will cover a large part of the sky (Fig. 7.1), and will take hours, instead of 2 min as it does now, to rise and set. Instead of being yellowish in color as it now appears, it will appear red, even redder than the setting Sun. Because of its color and size, astronomers call these old stars red giants.

The heat from this future Sun will evaporate the oceans and destroy almost all life forms on Earth. It will remain in this stage for several hundred million years. After that, the Sun will expand again. This time, it will engulf the Earth and its surface will approach the orbit of Mars. Its radiation output will increase to several thousand times the power that it has now. The astronomers call this stage the Asymptotic Giant Branch. We estimate that this stage will last about one million years.

The term asymptotic giant branch (AGB for short) is a terrible name, because it does not convey any meaning about its appearance, properties or its intrinsic nature. The term was given at a time when we did not have a good understanding of the nature of these stars. What they are, really, are stars that have gone through one phase of becoming redder and larger, and are doing the same thing again for the second time. As a member of the International Astronomical Union Working Group on Red Giant Stars, I proposed to have this name changed. Various suggestions were made by experts in the field but none of them is satisfactory. One particularly interesting term is "second-ascent-red-giant" suggested by American astronomer Robert Wing, which is a scientifically correct description. When "second-ascent-red-giant" is abbreviated to SARG, it has the meaning of coffin in German, very appropriate for a star that is just about to die.

S. Kwok, *Stardust*, Astronomers' Universe, DOI 10.1007/978-3-642-32802-2_7,
© Springer-Verlag Berlin Heidelberg 2013

Fig. 7.1 Earth's landscape when the Sun becomes a red giant.
This is an artist's conception of the Earth's landscape when the Sun becomes a red giant. The Sun would appear much brighter and larger than it is now

It is interesting to note that the element carbon, which is the basis of all organic compounds and life, is not made until the late life of a star. Our Sun has a total life span of about ten billion years, and yet carbon is not made until the last million or so years, during the AGB phase. If we scale the ten billion year lifespan of a star to a typical lifespan of a human being of 80 years, then the last million years correspond to the last 3 days of a human's life.

The other heavy elements come even later than carbon. After these elements are formed in the core of stars, they are brought to the surface through the process of convection. After that, these newly brewed chemical elements are blown away by a stellar wind. In the lives of humans, we are most active and lively when we are young. But for stars, they have a quiet life most of the time, and in the last hours of their lives, they get excited and do all kinds of weird things. These include changing color and shapes (see Chap. 8), making atoms and molecules (Chap. 9), and blowing smoke (see Chap. 10).

In Chap. 6, we talked about the discovery of stardust in old stars. More precisely, these small, solid particles are made by AGB stars. Contrary to the Greek belief that the stars are holy and the Earth is messy, we now learn that minerals, long supposed to be within the sole domain of the Earth, can also be found in stars. Even as late as the first half of the twentieth century, stars were still seen by astronomers as pure, in the sense that they only contain elements in the form of atoms and ions but nothing else. In fact, stars not only share the same chemical elements as the Earth, but may even have similar kinds of rocks.

It is fortunate that silicates and silicon carbides have emission features around the wavelength of 10 μm because the Earth's atmosphere is semi-transparent at this

Fig. 7.2 A view near the summit of Mauna Kea.
Although Hawaii has a reputation as a tropical paradise, the mountain of Mauna Kea is a desolate volcanic site. This view of Mauna Kea can easily be mistaken for a picture of the surface of lifeless Mars. Picture by the author

wavelength. Infrared astronomers refer to this gap as "windows of the atmosphere" as we can peek through this window to look at the infrared sky. In the other parts of the infrared, from wavelengths of about 5–1,000 μm, the atmosphere is almost completely opaque, due to absorption by water and oxygen molecules in the atmosphere. Any infrared radiation that is emitted by stars will not be able to penetrate the atmosphere and we will not be aware of its existence. To get around this problem, astronomers build infrared telescopes on high mountain tops to minimize the atmospheric absorption. For example, at the 4,207 m altitude of Mauna Kea (Fig. 7.2), the water content in the air is only about half of that at sea level, and infrared observations can be made through the 10 μm and other spectral "windows" in the atmosphere.

Mauna Kea is a dormant volcano on the Island of Hawaii. Although it is only 4,200 m above sea level, if measured from the ocean floor, it has a height of 10,200 m which can be considered the tallest mountain on Earth. It got its name (Mauna Kea means White Mountain in Hawaiian) because of its snow-covered peak. Mauna Kea was discovered to be a favorable infrared observing site by Gerard Kuiper of the University of Arizona in the early 1960s. In 1970, the University of Hawaii built the first telescope, a 2.2-m telescope on Mauna Kea, after securing permission from the Hawaii Department of Land and Natural Resources. Since that time, Mauna Kea has grown into the major astronomical observing site in the Northern Hemisphere, hosting many telescopes including the 3.6-m *Canada-France-Hawaii Telescope*, the 8.1-m *Gemini North Telescope*, the 8.2-m *Subaru Telescope*, and the twin 10-m *Keck Telescopes*.

However, there is a limit to how high a mountain on which one can build a telescope. The highest mountain is Mount Everest in the Himalaya Range at 8,848 m. Even at that altitude, the absorption of infrared light by water vapor is still

Fig. 7.3 The *Kuiper Airborne Observatory*.
The *Kuiper Airborne Observatory* was the first flying infrared telescope. The Observatory was converted from a C-140 cargo plane and carried a 0.9 m telescope. Photo credit: NASA

appreciable. There is also the additional problem of access. In order to construct a telescope, we need roads to transport construction materials and logistical considerations rule out many possible favorable sites for astronomical observations.

The other possibility is to put a telescope on an airplane. Commercial airliners routinely fly to altitudes of 10,000 m, and special aircraft, e.g. the U-2 spy plane can fly even higher. In 1974, NASA put a 0.9-m telescope on a C-140 transport plane and named it the *Kuiper Airborne Observatory (KAO)* in 1975 (Fig. 7.3). The *KAO* operated until 1995, and returned excellent quality data on the infrared view of the Universe.

An even better solution is to get above the Earth's atmosphere entirely. Since the first launch of an artificial satellite *Sputnik* in 1957, all kinds of military and commercial satellites have been put into Earth's orbit. It is only natural that astronomers would want to put an infrared telescope in space.

In 1982, the *Infrared Astronomical Satellite (IRAS)* was launched. The *IRAS* satellite was a joint venture between the U.S., U.K., and the Netherlands. The Dutch built the spacecraft, NASA contributed the telescope and the Delta 3910 rocket as the launch vehicle, and the British provided the ground station. The *IRAS* satellite carried a 0.6-m telescope and 700 l of liquid helium to cool the telescope and the instruments to under 5 °K in order to minimize background infrared radiation so that they would not interfere with the weak signals that were coming in from the distant stars.

Over the short 10-month' lifetime of *IRAS* before the coolant ran out, the telescope opened our eyes to the infrared universe, and completed a survey of 97 % of the sky in

Fig. 7.4 The *Canada-France-Hawaii Telescope.* The *Canada-France-Hawaii Telescope,* built in 1977, was among the first astronomical telescopes to be erected on the summit of Mauna Kea, Hawaii. Located on a ridge, it suffers the least from atmospheric turbulence that degrades the quality of pictures taken by optical telescopes. Photo credit: CFHT

the infrared. In addition to taking pictures, the *IRAS* satellite measured the brightness of 250,000 stars and galaxies at four infrared wavelengths. Among this quarter of a million objects were a number of new, previously unknown stars. These are stars that radiate only in the infrared but are extremely faint in the visible.

After the completion of the *IRAS* mission, Bruce Hrivnak of Valparaiso University in Indiana and I observed a large number of these newly discovered infrared stars at the *Canada-France-Hawaii Telescope* (Fig. 7.4) and the *United Kingdom Infrared Telescope* on Mauna Kea. Many of them turned out to be red giants heavily obscured by their own dust. They eject so much dust that these solid grains form a cocoon around the star and hide it from our view. The only way we learned that they are red giants is because they show the same silicate features as Mira Variables. We concluded as a red giant gets older, it sheds more and more stardust until it is completely covered by a dust cocoon. These infrared stars are the oldest red giants we know.

Also on board the *IRAS* satellite was a Dutch-built infrared spectrometer called the Low Resolution Spectrometer. In spite of its modest name, it was the first infrared spectrometer to be flown in space, and the first to do a spectroscopic survey of the sky in the infrared. At the conclusion of the mission, it obtained the spectra of over 50,000 sources. Analyses of the data found that over 3000 red giant stars show the emission feature due to silicate dust grains. Not only was the feature at 10 μm present, but so was an accompanying feature at 18 μm. There is no doubt that the manufacture of silicates is very common among old stars.

Carbon stars were not left out of the LRS survey. Although they are not as numerous as oxygen-rich stars, over 700 stars were found by the LRS survey to possess the 11.3 μm silicon carbide feature (Chap. 6). Again, some of them are heavily obscured and have no visible counterparts. Some of these obscured carbon stars have previously been detected by the AFGL rocket survey, and the best studied example was AFGL 3068. Using AFGL 3068 as a guide, Kevin Volk of the University of Calgary in Canada was able to find many more similarly obscured carbon stars. Because of their extreme red color, Kevin Volk named them extreme carbon stars.

The discovery of these extreme carbon and oxygen-rich stars is important for the understanding of stellar evolution. These oldest and most evolved stars were not observable by optical telescopes and we were therefore unable to study the very last stage of stellar evolution. The opening up of the infrared window gave us the first peek into the last stages of stellar evolution.

We now know that all the stars that manufacture stardust are AGB stars, a very short stage undergone by almost every star just before it dies. The infrared stars that were discovered by the *IRAS* satellite are not only the oldest of all AGB stars. Since stars in the AGB stage increase in brightness rapidly with age, these infrared stars are also among the brightest of all AGB stars. They radiate as much power as several thousand Suns and, in the most extreme cases, they can have intrinsic brightness more than ten thousand times that of the Sun. Yet in spite of their vast power and luminous output, they are invisible! They are shrouded by the dust of their own making. This dust cocoon absorbs all the visible light from the star and re-radiates the energy in the form of infrared light. So to be more precise, they are invisible only in visible light, but they shine very brightly in the infrared. By studying these newly discovered infrared stars, which we believe are older than any known red giant, we can get a glimpse of how stars behave when they get old.

The question of how stars die has fascinated astronomers for centuries. In contrast to the Greek scholars who believed that all celestial objects are immortal, astronomers since the nineteenth century have suspected that stars have a finite life cycle, with new stars being born as old stars die. In the twentieth century after we have ascertained that the Earth is billions of years old through evidence of isotopic dating, we came to the conclusion that stars (which our Sun is an example), must also have long lives. But how do stars change through their life time? Do they grow in size from youth to adulthood like animals on Earth? Do they age in appearance as they grow old? These are very difficult and interesting questions.

We know that stars come in different colors; some are red and some are blue. Our Sun is yellow. It is also known that in general blue stars are intrinsically brighter and red stars are generally faint. Since temperature is related to color (see Chap. 5) – blue stars are hotter and red stars are cooler – astronomers noted that there is a positive correlation between luminosity and temperature. If one were to plot these two quantities on a graph, stars seem to lie on a straight line, and this is called the "main sequence". In the early twentieth century, it was generally believed that stars evolve down the main sequence, from hot, bright, blue stars to faint, cool, red stars. By the mid-twentieth century, it was realized that this picture is incorrect. In fact, stars are born at a certain point on the main sequence depending on their mass, and once there, they tend to stay for a long time. The actual life time of a star is strongly dependent on its mass. A star like our Sun can live for 10 billion years, but a star like Vega can only live for 10 million years. Before they die, stars become redder and brighter and become a red giant. This is followed by another phase of increase in brightness and size in the AGB phase.

What happens after the AGB stage? Since a supernova explosion is such a spectacular event, it was thought that it represents the end of a star's life. While this is true for a very small minority of stars (less than 5 %), the majority of stars (including our own Sun) do not die in this way. It is amazing that a question as important as the cause of death of stars and the future of the Sun was unknown until the mid-twentieth century. How we learned about the answers to these questions is the topic of the next chapter.

A brief summary of this chapter
Old stars are able to make large quantities of dust and eject it into the Galaxy

Key words and concepts in this chapter
- Life cycles of stars
- The element carbon is synthesized by nuclear reactions near the end of a star's life
- Viewing of the infrared sky through atmospheric "windows"
- Astronomical observations from Mauna Kea
- Airborne observatories and astronomical satellites
- Old stars surrounded by dust cocoons

Questions to think about
1. Why should stars have a life cycle? Why do stars have to die? Why can't stars live forever as was believed by the ancient Greeks?
2. When people get old, they become weak and some shrink in size. An older person is much less energetic than a young person. Why do stars get bigger and brighter when they are old?
3. Some of the brightest stars we have in the Galaxy are old, red giant stars. However, due to the fact that they are surrounded by their own dust cocoon, they are faint in the visible. Since astronomers traditionally rely on visible light to survey the sky, is it possible that we have missed some of the brightest members of the Galaxy? How can we improve on this situation?
4. Our Sun is in its middle age now. Its size and brightness have been stable for five billion years. According to stellar evolution theory, the Sun is expected to remain the same for another five billion years. We are fortunate that the Sun has been stable for so long, giving life a chance to develop on Earth. Is this a coincidence? Some stars have much shorter lives. Do we expect life to be present in the planetary systems around those stars?

Chapter 8
Neon Signs in the Sky

The answer to what happens to a star after the red giant stage was found in an area of astronomy entirely unrelated to stellar astronomy, in objects called planetary nebulae. In spite of the name, planetary nebulae have nothing to do with planets. These objects were given the name because William Herschel thought that their greenish colors resembled those of the planet Uranus. Planetary nebulae were first discovered in the eighteenth century because of their nebulous appearance. They were mixed together with other non-stellar objects such as star clusters, galaxies, and nebulae from which new stars are born. Better photographic observations in the nineteenth century revealed that planetary nebulae are not a collection of stars (as star clusters and galaxies are), but are gaseous in nature, containing a single star in the middle. On August 29, 1864, William Huggins (1824–1910) observed the planetary nebula NGC 6543 (Fig. 8.1) in the constellation of Draco with a spectrograph and found its light output to be concentrated within a single bright emission line. This demonstrated clearly that this nebula is not a collection of stars, but is made of gaseous materials.

William Huggins was a wealthy Englishman who had a keen interest in astronomy. At the age of 30, he built a private telescope in Upper Lulse Hill, south of London. His interest in employing the new technique of spectroscopy led him to the discovery of the gaseous nature of nebulae and his position as a pioneer of modern astronomy. Instead of just recording the brightness, measuring positions, and taking pictures of celestial objects as was commonly practiced, astronomical spectroscopy allowed astronomers to study the physical construction of stars and nebulae. His marriage to Margaret Lindsay Murray in 1875 was the beginning of a collaborative effort on the spectroscopic observations of celestial objects, leading to the publication of *Atlas of Representative Spectra* in 1899, which received the Actonian Prize of the Royal Institution. Huggins was able to relate the lines he found in the spectra of astronomical sources to chemical elements on Earth, an accomplishment now recognized as the beginning of the discipline of astrophysics.

S. Kwok, *Stardust*, Astronomers' Universe, DOI 10.1007/978-3-642-32802-2_8,
© Springer-Verlag Berlin Heidelberg 2013

Fig. 8.1 NGC 6543, the
Cat's Eye Nebula.
The Cat's Eye Nebula in the
constellation Draco is one of
the most dusty planetary
nebulae. This *Hubble Space
Telescope* image of the object
shows a bright central star.
The different colors are the
result of emission from
different atomic emissions.
This planetary nebula also
shows strong infrared
emission from solid particles
as in the case of NGC 7027.
Image created by the author
based on data from the
Hubble Space Telescope
archive

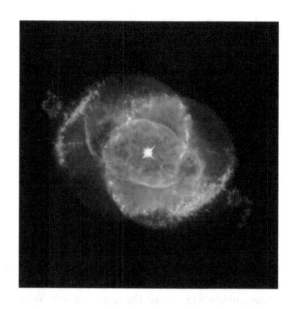

A Brief Course on Stellar Evolution

Stars do not live forever. Like human beings, they have finite lifetimes. Every year, new stars are born and old stars die in the Milky Way Galaxy. The evolution of stars is primarily driven by a single factor, the availability of nuclear fuel and the necessity of adjusting stellar structure to take advantage of available fuel. Our Sun, which is typical of the majority of stars in our Galaxy, relies on hydrogen for nuclear fuel for the majority of its life (about 10 billion years). The ash of burning hydrogen is helium and helium gradually replaces hydrogen to form the dominant matter of the core. After all the hydrogen in the Sun's core is used up, the core of the Sun has to contract and its envelope has to expand to make use of the hydrogen in the layers outside the core. This is when the Sun enters the red giant phase. During this phase, the size of the Sun increases by a hundred fold, and its luminosity brightens by several hundred times. After another few hundred million years, the core of the Sun is hot enough to begin fusion of helium. The product of helium fusion is carbon and oxygen. When the Sun enters the phase where its core is composed of carbon and oxygen, it will expand again, converting hydrogen and helium outside the core to helium and carbon respectively. This phase is called the asymptotic giant branch (AGB, Chap. 4).

The spectroscopic studies of Huggins led to a surge of interest in the study of planetary nebulae, which in turn opened up our understanding of the death of stars. Planetary nebulae were first found as nebulous objects in the sky as the result of visual observations and then photography. With the invention of spectroscopy in the late nineteenth century, planetary nebulae became popular objects for astronomical observations because of their bright emission lines, making them ideal for spectroscopic observations. The development of atomic physics in the early twentieth century allowed us to use this new-found knowledge to identify the chemical elements in the nebulae, and determine their relative abundance.

In the early twentieth century, astronomers were puzzled by a bright green line which they could not associate with any known elements on Earth. The name of "nebulium" was coined in the belief that they had discovered a new element not found on Earth. In 1926, Ira S. Bowen of the California Institute of Technology was able to identify this line as due to the common element oxygen. The reason why this line is not seen in terrestrial laboratories is that the high density atmosphere of the Earth suppresses this line. Only in a low-density environment like a planetary nebula would such an atomic transition take place. Identification of many other similar "forbidden lines" followed. Such discoveries were only possible through a marriage of knowledge of observational astronomy and laboratory and theoretical studies of atoms. We now have access to conditions not previously possible for the study of atomic physics and many consider the study of planetary nebulae to be responsible for the transformation of the traditional discipline of astronomy to the modern discipline of astrophysics.

One of the towering figures who were responsible for this transformation was Lawrence Aller (1913–2003). Aller was born in Tacoma, Washington, USA and had a very tough childhood. At the age of 15, he found the book *Astronomy* by Russell, Dugan, and Stewart in the Seattle Public Library and learned about the exciting development of early twentieth century astronomy. His father was very much against his interest in astronomy, considering it "impractical". He was dragged out of school and put into a mining camp. He studied science on his own and through the help of Donald H. Menzel (who was on leave from the *Lick Observatory*), got admitted to University of California at Berkeley at the age of 19 as a special student. After a classical education in Berkeley, he wanted to follow Menzel to Harvard. He went there in 1937 with a minimum scholarship that paid his tuition and had to really struggle to make ends meet. For his thesis work, he performed spectroscopic observations of planetary nebulae under the guidance of Menzel. During the War, he worked in the Radiation Laboratory in Berkeley on the separation of the uranium isotopes 235 and 238. He knew that this work had to do with the atomic bomb, but in May 1945, 3 months before the dropping of the atomic bombs in Japan, half of the staff of the Lab was sacked, Aller among them. After the war, Aller was professor at the University of Indiana, University of Michigan, and later at the University of California at Los Angeles. Armed with then new knowledge of atomic physics, Aller was able to interpret the spectra of planetary nebulae and derived the physical conditions and chemical compositions of the objects. He later wrote several textbooks in the 1950s and 1960s which laid the foundation of modern astrophysics. His encyclopedic knowledge of astronomy was just amazing. Aller remained active in research until he died at the age of 90.

With modern telescopes, thousands of emission lines from dozens of atomic elements from common chemical elements such as hydrogen, carbon, oxygen, sulfur, neon, iron, to rare elements such as selenium, bromine, krypton, rubidium, strontium, xenon, can be observed and identified in the optical spectra of planetary nebulae. These detections demonstrate clearly that chemistry is universal in the Universe. Since atoms radiate differently under different physical conditions, observations of different atomic lines allow us to determine the temperature and density of the nebulae. In spite of their large distances from us, we intimately know the conditions of these objects. Through remote observations, we can probe the conditions of distant objects without having to actually go there. Such is the power of modern astronomy.

In spite of all the knowledge of planetary nebulae gained through observations, the true nature of planetary nebulae in the scheme of stellar evolution was not known until the mid-twentieth century. Astronomers have struggled to understand how planetary nebulae fit into the overall life cycle of stars (see box "The Life Cycle of Stars"). A first step in finding an answer to this question took place when the Russian astronomer Iosif Shklovsky determined that they are very old stars just about to die. This theoretical understanding opened up a fascinating area of research in the second half of the twentieth century.

The Life Cycle of Stars

Most of the synthesis of heavier chemical elements occurs during the AGB phase. At this time, the Sun will be a few hundred times larger than it is now and several thousand times brighter. It also begins to develop a strong stellar wind, ejecting part of its outer atmosphere into space. When the wind completely removes the atmosphere of a star, the core is exposed and we enter the phase of planetary nebulae (Chap. 8).

Among the materials ejected in the stellar winds are carbon atoms, freshly synthesized in the core and taken up to the surface through convective processes. Molecules and solids begin to form in the stellar wind, leading us to call the old AGB stars "molecular factories in the sky" (Chap. 7).

Planetary nebulae expand and gradually disperse themselves into the interstellar medium. They are only visible for about 30,000 years. After that the nebulae are too faint to be seen and the only remnant of the star that can be observed is the core. We call these objects "white dwarfs" as they are hot (and therefore appear white) and they are no longer luminous as their nuclear fuel is running out. The gradually fading of a white dwarf represents the end of typical stars in our Galaxy. When the shining moment of a star ends, the Galaxy becomes its graveyard. At the same time, the gas and dust ejected earlier may aggregate together elsewhere in the Galaxy and become the birth sites of another generation of stars.

We now know that planetary nebulae are gaseous nebulae that are ejected by stars when they are near the end of their lives. They are the immediate descendants

Fig. 8.2 *Spitzer Space Telescope* image of the Ring Nebula.
In this infrared image, the Ring Nebula (NGC 6720) shows a flower-like structure with petals outside the main nebula. For 200 years, only the inner part of the Ring Nebula was known to astronomers. This color image is constructed by combining infrared images obtained at different wavelengths. The white dots are background stars. A galaxy can be seen on the top right side of the picture. Image credit: C.H. Hsia

of the oxygen and carbon-rich red giant stars that we talked about earlier. The gaseous and solid materials inside a planetary nebula originate from the ejecta of these red giants. What make planetary nebulae fascinating are their beautiful colors and magnificent shapes. Unlike stars, which radiate in all colors and therefore appear white, planetary nebulae emit strongly at specific colors due to atomic emissions. Their rich colors have been referred to as the glorious firework displays just before a star's death. Because the radiation mechanism of planetary nebulae is similar to that of neon lights, I have called them "neon signs in the sky". It is not an exaggeration to say that the many beautiful pictures of planetary nebulae taken by the *Hubble Space Telescope* have motivated or tempted many young people to contemplate a career in astronomy.

The most well-known planetary nebula is the Ring Nebula in the constellation of Cygnus (Fig. 8.2), which can be easily seen with a small telescope in the summer sky. For the 200 years since its discovery in 1779, the Ring Nebula has been famous for its ring-like (therefore its name) structure. However, recent observations with more powerful telescopes have revealed that the Ring Nebula has layers of flower-like patterns outside the ring. This has opened up all kinds of new speculations and theories about its true structure. While many planetary nebulae (e.g. NGC 6302 and NGC 6537, Fig. 8.4) have butterfly-like structures, many others have simple elliptical shapes. The question of the origin of the diversity of shapes is one of the most debated questions in astronomy circles. Something mystical must have happened to the stars to turn its once spherical shape into a butterfly.

Fig. 8.3 *Hubble Space Telescope* images of five proto-planetary nebulae.
The 5 proto-planetary nebulae are (clockwise from *top right*): the Water Lily Nebula, the Spindle
Nebula, the Cotton Candy Nebula, the Silkworm Nebula, and the Walnut Nebula (*center*). These
nebulae were first discovered through their infrared colors, and then followed up by imaging
observations in the optical with the *Hubble Space Telescope*

Planetary nebulae are interesting not only because of their colors, but also their appearance. As we all know, the Earth is round, the Moon is round, the Sun and the stars are round. There is a good reason for roundness as it is the shape desired by a body held together by gravity. However, planetary nebulae have many different shapes, but they are rarely round. Many planetary nebulae have butterfly shapes, and some are like flowers with several petals. How they acquire these shapes is still a subject of intense debate in the scientific circles. What we do know is that soon after a star leaves the red giant stage, it undergoes a metamorphosis. As the reader may recall, at the end of the red giant stage, stars wrap themselves in a cocoon of dust and gas. As soon as the cocoon is woven, another energetic process from the star breaks the cocoon and out emerges a butterfly. I was greatly intrigued by this process. I wanted to identify objects that are in transition from red giants to planetary nebulae and study them so that we can get some clues to this transformation. Since we knew that the stellar cocoons are infrared objects, we searched stars in the *IRAS* catalogue that had the infrared colors that we expected. Guided by theoretical models developed by Kevin Volk, Bruce Hrivnak and I discovered about 30 of these transition objects. Since these objects are expected to evolve into planetary nebulae, they were given the name of proto-planetary nebulae.

Figure 8.3 shows some examples of proto-planetary nebulae that we discovered. Even at this stage of infancy, they already possess butterfly shapes. Proto-planetary nebulae are important not only because they tell us about the process of morphological transformation, but they also hold the clue to the synthesis of organic compounds by stars, as we shall learn in Chap. 10.

Another unique aspect of planetary nebulae is their richness in composition. Matter in our world can be in several different states. Rocks are solids, water is a liquid, and air is in gaseous form. Sometimes molecules are broken up into individual atoms. When some energetic processes hit atoms, they may lose some of the electrons and be left in an ionic (or plasma) state. Planetary nebulae are among the few celestial objects that contain almost all the states of matter known: ions, atoms, molecules, and solids. Since each state of matter can have its own unique way of radiating, this rich mix of matter makes planetary nebulae radiate throughout the electromagnetic spectrum from the long wavelengths of radio (with wavelengths of centimeters or longer) to the energetic X-rays (with wavelengths fractions of nanometers, see Appendix B). The ability to radiate over such a wide range of colors is rare among astronomical objects.

The temperatures inside planetary nebulae range from a cold 100 K for the solids, to about 10,000 degrees for the ions, and to millions of degrees in the shocked heated bubbles that generate the X-rays. Such a wide range of conditions makes them a paradise for physicists and chemists who can use planetary nebulae as a laboratory to study all kinds of exotic physical and chemical processes not possible on Earth.

As the last hurrah in the life of a star, the planetary nebula is also believed to be a final passage that our Sun will go through in another 5 billion years. After the red giant stage when the bloated Sun has evaporated or swallowed the nearby planets Mercury, Venus, Earth, or even Mars, streams of gas and dust will emerge from the Sun's surface, creating a huge cocoon. After almost the entire atmosphere of the Sun has been ejected, the speed of the wind increases by a hundred fold and this new fast wind piles up the previously ejected cocoon into a giant shell. The removal of the outer layers of the giant Sun exposes the hot core, and the bright ultraviolet light from the hot core lights up the atoms in the shell and turns the shell into a collection of colorful neon signs. This fabulous display, however, will last only a couple tens of thousands of years. Afterwards, the nebula disperses into the general interstellar medium. At the same time, the central star, having used up all its energy in this display, dims and fades into obscurity ending in a stage known as a white dwarf.

In common folklore, star deaths are often associated with supernovae. Supernovae are exploding stars resulting in sudden increase in brightness to more than a billion times the brightness of the Sun. In fact, supernovae are rare. The Milky Way galaxy has about 100 billion stars. Every year, about 1 star will die and become a planetary nebula in the Galaxy. The rate of supernova explosion is estimated to be about one per century. The last supernova explosion actually witnessed in the Galaxy was the supernova observed by Johannes Kepler in 1604. The one before that occurred in 1572, observed by Tycho Brahe. There has been no supernova seen in the last 400 years although we suspect that some might have happened on the other side of the Galaxy but their light was obscured by intervening dust between the supernovae and the Sun.

Supernovae events are rare because most of the stars in the Galaxy have low mass, and only stars with masses 10 times that of the Sun will become supernovae. All others (about 95 %) will go through the spectacular, but not explosive, planetary nebulae phase.

Up to the early 1970s, planetary nebulae have been primarily studied with optical telescopes. When infrared astronomy was first developed (Chap. 5), planetary nebulae were surprisingly found to be bright infrared objects. These infrared properties led me to believe that planetary nebulae are related to the infrared-bright red giant stars (Chap. 7) and the infrared light from planetary nebulae may be the result of emission from remnants of stardust left over from the preceding evolutionary stage, the AGB phase. If this is the case, then infrared emission must be common among planetary nebulae. This prediction was confirmed after the all-sky survey by the *IRAS* mission (Chap. 7), where over 1000 planetary nebulae were detected. These results show that planetary nebulae contain a large amount of dust. The question is: are these dust particles all inherited from their parent stars, or are they freshly made inside the planetary nebulae themselves?

The success of *IRAS* really whetted the appetites of astronomers. They wanted to build a longer-lasting infrared telescope in space which would be capable of targeting specific objects. One of the greatest technical challenges is that in order for the weak stellar infrared signals to be detected, the background infrared radiation has to be minimized. This requires cooling the telescope and the instruments to very low temperatures. Usually this is achieved with the use of liquid helium as a cooling agent. Since liquid helium evaporates with time, the supply of liquid helium sets a limit on the lifetime of the usefulness of the telescope.

The product was the *Infrared Space Observatory (ISO)*, a mission led by the *European Space Agency* with contributions from the U.S.A. and Japan. The *ISO* satellite was equipped with a 0.6-m telescope and was cooled by a tank of 2,300 l of liquid helium. This 2.3-ton spacecraft was successfully launched from an Arianne IV rocket on November 17, 1995 from Europe's space port in Kourou French Guiana near the Caribbean coast of South America. Like *IRAS* before it, it was a remarkable success. Before it ran out of coolant in April 1997, *ISO* made 30,000 observations of stars and galaxies.

ISO also carried much more advanced and sophisticated infrared instruments than *IRAS*. In addition to an infrared camera, it had two spectrometers, one called the Short Wavelength Spectrometer (SWS) built by a team led by Thijs de Graaus of the Laboratory for Space Research in Groningen, the Netherlands, and the other called Long Wavelength Spectrometer (LWS) built by a team led by Peter Clegg of the Queen Mary College of the University of London in England. These two spectrometers could make much more sensitive and precise measurements than the *IRAS LRS*. *ISO* observations of planetary nebulae soon identified many atomic lines in the infrared, emitted by the elements oxygen, nitrogen, sulfur, neon, magnesium, silicon, argon, and many others. These measurements allowed astronomers to determine many atomic parameters much more accurately than can be done in the laboratory and planetary nebulae rightfully earned their place as the laboratory in the sky.

Although planetary nebulae have been mostly known as objects of visual beauty, it was only after the *IRAS* and *ISO* missions that we learned that gas is not the only component in the nebulae, but there are molecules and dust as well. The solid-state

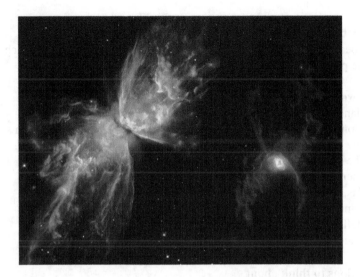

Fig. 8.4 *Hubble Space Telescope* images of NGC 6302 (left) and NGC 6537 (right).
Hubble Space Telescope images of the southern planetary nebulae NGC 6302 and NGC 6537.
Both nebulae have remarkable butterfly-like structures. The *ISO* measurements are made toward
the waist of the objects, where we believe most of the solid state minerals lie. The gas in the
nebulae has a temperature of about 10,000 K and the central star of NGC 6302 is one of the hottest
star known (about 250,000 degrees). Image credit: NGC 6302: NASA and ESA. Image of NGC
6537 by the author based on data from the *Hubble Space Telescope* archive

dust in the nebulae revealed itself in the form of infrared light. Planetary nebulae are
found to be strong infrared sources, with as much as one third of their energy being
emitted in the infrared. Light from the central star is absorbed by these dust particles
and in the process are heated up to about 100 K. These "warm" (relative to the
extreme cold condition of general interstellar space) dust then get rid of their energy
by radiating in the infrared, mostly in the mid-infrared at the wavelength of about
30 μm.

The infrared spectrometers on board *ISO* allowed us to study these dust particles in
detail. One of the discoveries of the *ISO* mission was the detection of crystalline
silicates and carbonates, the main constituents of common rocks and limestones
respectively. Silicates are the most abundant group of minerals in the Earth's crust.
Unambiguous detections of forsterite (Mg_2SiO_4) and enstatite ($MgSiO_3$) have been
made in a number of planetary nebulae, including NGC 6302 and NGC 6537 (Fig. 8.4).

However, minerals are not the only chemical constituents of dust in planetary
nebulae. *ISO* found that organic grains are also widely present. This raised all kinds
of new questions, for example: what kind of organics are they? How were they
made? Since planetary nebulae are in a rapidly evolving state, how can complex
molecules be made in such volatile conditions? The discovery of stellar organics is
one of the most surprising events in modern astronomy. We will come back to these
questions in Chap. 10.

A brief summary of this chapter
The beautiful and magnificent planetary nebulae are dying stars which are active producers of stardust.

Key words and concepts in this chapter
- Planetary nebulae as neon signs in the sky
- Atomic spectroscopy and the identification of chemical elements in stars
- The beginning of astrophysics
- The mysterious green line
- Chemistry is universal in the Universe
- Gas, molecules, and solids all co-exist in planetary nebulae
- Planetary nebulae as the end stage of stellar evolution
- Stellar metamorphosis and proto-planetary nebulae
- Planetary nebulae as infrared objects

Questions to think about
1. Have you thought about why the Sun and the Earth are round? What are the physical principles that cause the Sun and the Earth to have taken a round shape?
2. Why do the principles that hold stars round not apply to planetary nebulae? Can you think of some physical processes that can create butterfly shapes, as commonly seen in planetary nebulae?
3. For most people, their everyday experience gives them the impression that the Earth is flat. But the Greeks more than 2000 years ago knew that the Earth is round, and were able to accurately determine the size of the spherical Earth. How did they do that?
4. The different morphological appearance of planetary nebulae has led astronomers to give them nick names such as "The Owl", "The Dumbbell", "Cat's Eye", "Eskimo" and many others. For planetary nebulae discovered in modern times, the right of naming is usually reserved for the people who discover them; although this is not strictly enforced as there are cases where people naming objects already known. Do you think there ought to be a set of rules for naming, or it should be left open?
5. Planetary nebulae are thousands of light years away from us, yet with remote astronomical observations, astronomers are able to determine the temperature and density within the nebulae without making any in-situ measurements. How is this possible?
6. Planetary nebula is a transient phenomenon associated with the death of stars. Why do stars have a finite life time with birth and death, and are not ever lasting as the ancients believed?
7. The element helium was found on the Sun before it was found on Earth. Many of the chemical elements found on Earth are also found in stars and nebulae. Is this star-Earth connection surprising? What is the philosophical implication of this cosmic connection?

Chapter 9
Stars as Molecular Factories

A micrometer-size particle of stardust is made of billions of atoms. In the case of silicates, the atoms are oxygen, silicon, iron and magnesium. How did so many atoms aggregate themselves into a solid? On Earth, when atoms meet, they form molecules. Molecules are simple entities that consist of two or more atoms. The air we breathe contains primarily oxygen (O_2) and nitrogen (N_2) molecules, which are made of two oxygen and two nitrogen atoms, respectively. Water, one of the most common molecules on Earth, is made of two hydrogen atoms and one oxygen atom. The water molecule can be in a gaseous form (steam), liquid (water), or solid (ice). The ice crystal contains many water molecules arranged in fixed geometric patterns. If we think about the relationship between water and ice, we must realize that the first step to form ice is to have water molecules. This is also true for stardust. The first step to form silicates is to have simple oxide molecules. We can imagine that the atoms in the stellar atmosphere collide with each other, first forming simple gas-phase molecules which later aggregate into larger solids. Since we see stardust through the infrared radiation it emits, there must also be simple molecules in these infrared stars. The question is: how do we detect them?

The realization of the existence of molecules has not been an easy process. It all started with an attempt to understand the question "what is in the air? The Earth's atmosphere is filled with air that we breathe. We know there is something in the air because we cannot survive without it. When humans or animals are deprived of air, they die within minutes. So air must contain something that we need. We also can feel the wind when air moves. Since the discovery of fire in ancient times, human beings knew that air was needed for lighting a fire. What is that something? Is the ingredient needed for life the same as the ingredient needed to light a fire? Are there other parts of air that are needed for something else, or serve no purpose at all? These simple, fundamental questions have been on the minds of curious, thinking people for hundreds, if not thousands of years.

By the early eighteenth century, the ingredient necessary for combustion was assumed to be a hypothetical substance in combustible objects called "phlogiston" (a word from Greek, meaning "burnt"). It was not until 1772 that

S. Kwok, *Stardust*, Astronomers' Universe, DOI 10.1007/978-3-642-32802-2_9,
© Springer-Verlag Berlin Heidelberg 2013

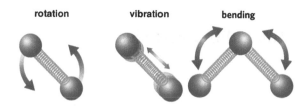

Fig. 9.1 How molecules are detected.
A 2-atom molecule can rotate and stretch and a 3-atom molecule (such as CO_2) can rotate, stretch and bend in different ways. When it undergoes these motions, the molecule radiates in specific frequencies therefore allowing astronomers to detect its presence

the Swedish chemist Carl Wilhelm Scheele showed that air contains two different kinds of gases: one supports combustion (which we now know as oxygen) and the other prevents it (which we now know to be nitrogen). The role of oxygen in combustion and respiration was later confirmed by the French chemist Antonine Lavoisier, who gave the gas its name.

In the early nineteenth century, scientists began to suspect that air was made of many small particles (which we now call molecules) which in turn were made of combinations of distinct fundamental particles (which we now call atoms). This view of the molecular composition of gas was proposed by an Italian, Amedeo Avogadro, who correctly stated that water is made of two hydrogen atoms and one oxygen atom, and ammonia is made of three hydrogen atoms and one nitrogen atom. However, Avogadro's idea did not gain much acceptance during his lifetime. This could be partly attributed to the fact that he stayed mainly in Turin, Italy, and did not interact with other prominent scientists. It was only after his death in 1856 that his molecular hypothesis gradually gained recognition in scientific circles. However, we should not be too harsh on his scientific contemporaries because we cannot see and touch air, and the arrival at the correct conclusions on the composition of air required careful measurements in the laboratory.

Returning to the question of detecting molecules in space, if we cannot see the molecules right in front of our nose, how can we see molecules in a star hundreds of light years away? Indeed this is a formidable task.

Although molecules do not emit visible light, they do radiate in the infrared and in the radio. Since molecules are made of two or more atoms, they can twist and turn in addition to moving in a straight line through space. Let us take the simple molecule carbon monoxide as an example. Carbon monoxide is made of two atoms: carbon and oxygen. We can imagine the structure of this molecule as a dumbbell, with the two atoms in the form of balls connected by a rod. When we twirl the dumbbell, we say that the molecule is rotating. If the rod is not rigid, the oxygen and carbon atoms can also move toward or away from each other, and we say the molecule is stretching (Fig. 9.1). In the early twentieth century, physicists realized that when molecules rotate or stretch, they emit radio (or more precisely, microwave) and infrared signals, respectively. By measuring these frequencies in the laboratory, physicists and chemists are able to compile extensive lists of frequencies tied to specific molecules. This technique is known as molecular spectroscopy. One of the towering figures in this field is Gerhard Herzberg (1904–1999), who won a Nobel Prize in 1971 for his work in molecular spectroscopy.

Molecules vibrate and rotate under any temperature, even in space at temperatures as low as tens of degrees Kelvin. In doing so, they emit radio signals of unique frequencies not unlike radio stations having unique broadcasting frequencies. If we mount a radio receiver onto an antenna and point the antenna at a celestial source, we can search for signals that originate from molecules in space. Unfortunately, most simple molecules (which are likely to be the most abundant) emit radio signals at very high frequencies. Typical frequencies corresponding to the rotational signatures of 2, 3, or 4 atom molecules occur at about 100 GHz, or about several hundred times higher than the frequencies used for commercial television broadcasts. The radio frequencies used by FM broadcasts are in the range of 87 to 108 MHz, whereas television broadcasts use the VHF (Very High Frequency) band, which is in the range of 30 to 300 MHz. By the early twenty-first century, commercial radio frequencies have extended to higher frequencies; for example mobile phones use radio frequencies around 900 or 1,800 MHz.[1]

While commercial users were satisfied with the use of radio technology below 1,000 MHz (or 1 GHz), astronomers became the driving force behind the development of high-frequency radio receivers in the late 1960s. Robert Wilson and Arnold Penzias, two physicists working at Bell Telephone Laboratories embarked on exactly this venture. They had just discovered the cosmic background radiation, a radio hiss noise left over from the beginning of the Universe. This discovery led to the award of the Nobel Prize to them in 1978. Instead of stopping with this momentous achievement, they led a group at Bell Labs to build a high-frequency radio receiver for astronomical use. By mounting this newly built receiver onto the *National Radio Astronomy Observatory*'s 12-m telescope at Kitt Peak, Arizona (Fig. 9.2), they were able to identify the rotational signatures of the carbon monoxide (CO) molecule. In 1970, they found the first signs of these molecules in the Orion Nebula, an interstellar cloud where new stars are currently being born.

While the stellar origin of the element carbon has been known since the 1950s, it was only after the discovery of infrared stars such as CW Leo (Chap. 5) that we realized these old carbon stars are also capable of turning carbon atoms into molecules. Using the same receiver, Wilson and Penzias, working with Phil Solomon of the University of Minnesota, found that the infrared star CW Leo was producing and ejecting into space trillions of tons of carbon monoxide molecules every second! Although this star was unknown to astronomers until 1969, we now know that it is only one of thousands of old stars making molecules and polluting the Galaxy at this moment.

---

[1] For historical reasons, radio technology uses frequency as the unit of radiation whereas astronomers use wavelengths. Both frequency and wavelengths are ways to describe the color of electromagnetic radiation. For a conversion between these two unit systems, see Appendix II.

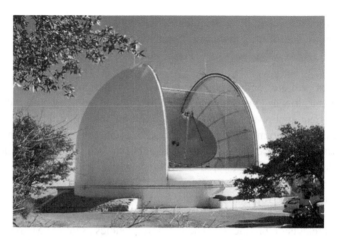

Fig. 9.2 The 12 m telescope at Kitt Peak.
The 12 m telescope at Kitt Peak, Arizona, built by the *National Radio Astronomy Observatory* was the telescope used to discover the molecule carbon monoxide in space. For 30 years, it was the workhorse for the study of interstellar molecules until it was closed by *NRAO* in 2003. Its operation was taken over by the University of Arizona, and is now part of the *Arizona Radio Observatory*. Photo credit: NRAO

Technology and Astronomy

Because stars and galaxies are far away, the signals we get from them are often very weak. The need to observe celestial objects at ever larger distances has driven the development of new technology. In the early 1980s, astronomers were among the first to adopt the use of charge coupled device (CCD) cameras as imaging devices because of their sensitivity to light. Their linear response from faint to bright light also offers CCD an advantage over photographic plates. Now CCD cameras are so common that they have completely replaced film as a means of leisure photography. CCD cameras can be built so small that they even fit into cellular phones.

Although radio communication was first developed for wireless communication and broadcasting, commercial use of radio frequencies had been in the MHz range. The development of radar technology in World War II pushed frequency use to the GHz (or cm in wavelength) range as higher frequency allowed the detection of smaller objects. Since molecular transitions occur at much higher radio frequencies, astronomical needs were the main driver for the development of high frequency radio receivers. The building of the millimeter-wave (100 GHz) receiver in 1970 to search for the rotational line of CO was such an example. This was over a thousand times higher than the commercial frequencies used for TV and radio at the time. By the beginning of the twenty-first century, cellular phone frequencies had gone up to 2 GHz, and radio receivers used for astronomy had gone up to 500 GHz. It is not uncommon for astronomical technology to be decades ahead of the commercial market.

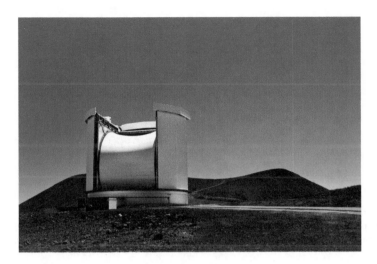

Fig. 9.3 The *James-Clerk-Maxwell Telescope* on Mauna Kea.
The *James-Clerk-Maxwell Telescope* on the 4,200 m summit of Mauna Kea, is an example of the modern radio telescopes capable of detecting millimeter-wave emissions from molecules. Photo credit: Joint Astronomy Center

The success of finding molecules in space has motivated many countries to construct telescopes operating at millimeter and submillimeter wavelengths. In order to minimize the obstruction effects of the atmosphere, most of these telescopes are built on high mountain sites. The most popular sites are on Mauna Kea, Hawaii (Fig. 7.2) and on the mountains of Chile. One example of such telescopes is the *James-Clerk-Maxwell Telescope (JCMT)* on Mauna Kea shown in Fig. 9.3. The *JCMT* is a 15-m telescope jointly developed by the United Kingdom, Canada, and the Netherlands and was one of the first large submillimeter-wave telescopes operating on a high mountain site. Other examples include the 30-m *Institut de Radioastronomie Millimétrique (IRAM)* telescope located on Pico Veleta in the Spanish Sierra Nevada at an altitude of 2,850 m, the 10-m *Atacama Submillimeter Telescope Experiment (ASTE)* telescope located at the Atacama desert, Chile of altitude 4,860 m, and the 12-m *Atacama Pathfinder Experiment (APEX)* telescope located at Llano de Chajnantor (5,105 m altitude) in Northern Chile.

Further observations of these carbon stars have led to the detection of over 60 molecules, including organic species such as polyacetylene radicals (C_6H, C_8H), cyanopolyynes (HCN, HC_3N, up to HC_9N), and sulfuretted chains (C_2S, C_3S). The largest molecule HC_9N has an atomic weight of 123, almost twice the molecular weight of the simplest amino acid glycine.

Some of the molecules found in CW Leo do not occur naturally on Earth. Some molecules are not stable and only live for a very short time in our high-density atmosphere. Since their rotational frequencies are not known in advance from measurements in the laboratory, astronomers identify these molecules by

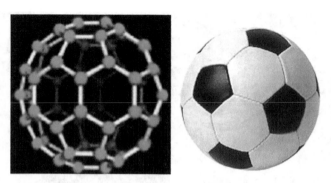

Fig. 9.4 Chemical structure of fullerene.
The fullerene (C_{60}) molecule (*left*) consists of 60 carbon atoms (shown as green balls) arranged in adjacent pentagons (5-carbon rings) and hexagons (6-carbon rings). Its structure is similar to the soccer ball (*right*) where the pentagons are shown in *black* and the hexagons are in *white*

analyzing the patterns of unidentified lines in their telescope spectrum. Using theoretical knowledge about the structure of molecules, astronomers can guess from the observed patterns which molecule is likely to be responsible. This is often later confirmed by laboratory spectroscopists who create these unstable molecules using specialized techniques. The study of these new species has led to the development of a new scientific discipline called astrochemistry.

Harry Kroto, a chemist at the University of Sussex in the United Kingdom, was interested in simulating the production of long-chain carbon molecules in space. In the late 1970s, Kroto frequently visited the Herzberg Institute of Astrophysics in Ottawa, Canada. Radio astronomers in Ottawa had been detecting long, linear, chain-like carbon molecules in interstellar clouds. How these very large molecules can be synthesized in space is a very interesting chemistry question. When I joined the Herzberg Institute in 1978, I was able to interest the group to observe old stars such as CW Leo, and indeed similar long chain molecules were detected there. The environment of these old stars is much better understood than that of interstellar clouds, and therefore a much better controlled laboratory for chemical studies.

Kroto got his colleagues Robert Curl and Richard Smalley at Rice University in Houston, Texas, to try to synthesize long-chain carbon molecules in the laboratory. Instead of long carbon chains, they discovered that 60 carbon atoms can group together in the form of a closed sphere. This molecule has 12 five-membered and 20 six-membered carbon rings curved to form a soccer-ball-like structure (Fig. 9.4). They derived this structure with inspiration from the structure of geodesic dome, an architectural structure based on a combination of geometric patterns curved together to form a closed sphere. Since the concept of geodesic dome was popularized by Buckminster Fuller (1895–1983), Harry Kroto named the C_{60} molecule Buckminsterfullerene (fullerene for short).

Although astronomers rely on physicists and chemists for information on the search for interstellar molecules, astrochemistry can also lend itself to the advancement of fundamental chemistry. The discovery of fullerene (C_{60}) as a new form of carbon beyond the known forms of graphite and diamond is such an example. This discovery of fullerene won Kroto, Curl, and Smalley the Nobel Prize in chemistry in 1996.

Diamonds and graphite have been known for 2000 years, the former as a gem and the latter as a coloring agent in ink and dyes. They were identified as pure forms of carbon in the nineteenth century. The discovery of a new form of carbon was therefore very exciting. Immediately after the laboratory synthesis of C_{60}, scientists raced to discover this molecule in space. They believed it would likely survive in space due to its remarkably stable structure, and the search was driven by Kroto, whose laboratory work was motivated by astronomy in the first place. Astronomers performed the first searches by looking for features of the molecule in the visible, as optical telescopes had the most advanced devices for spectroscopy. All these searches turned up nothing. When infrared spectroscopy became available through infrared satellites, interest turned to the infrared, where the molecule is known to have 4 bright bands. In spite of the high spectral resolution capabilities of the *ISO* satellite, no C_{60} was definitely detected.

It took 25 years after C_{60}'s laboratory synthesis for the molecule to be found in space. The *Spitzer Space Telescope* was launched in August 25, 2003. Although it does not have the spectral resolution of *ISO*, it is much more sensitive and therefore capable of detecting much weaker infrared signals. Over its lifetime of 5 years, *Spitzer* has taken many spectra of planetary nebulae. A spectrum of the planetary nebula Tc-1 was taken in 2005 but not much attention was paid to it. In 2010, Jan Cami, Jeronimo Bernard-Salas, and Els Peeters extracted the spectrum of this object from the *Spitzer* data archive and, to their surprise found the long-sought infrared features of C_{60} prominently displayed in the spectrum. Their discovery was published in the magazine *Science* and made headline news around the world.

Immediately after Cami's announcement, many astronomical detections of C_{60} followed. Arturo Manchado and Anibal García-Hernández of the Institute of Astrophysics of the Canary Islands discovered C_{60} in planetary nebulae, Kris Sellgren of the University of Ohio detected C_{60} in reflection nebulae, and Yong Zhang and I detected the molecule in a proto-planetary nebula. The detection of C_{60} and other molecules confirmed that old stars are extremely efficient molecular factories.

There is evidence that these stellar fullerenes have reached the Solar System. Analysis of the Allende meteorite showed that it contains C_{60}, and C_{70}, as well as the higher members of the fullerene family of molecules. These include C_{74}, C_{76}, C_{78}, C_{84}, and possibly extending to C_{100} or C_{250}. These detections suggest that these stellar molecules have traveled through the interstellar medium to our Solar System and were accreted into meteorites and delivered to Earth. Indeed fullerenes have been found in impact craters, further supporting the idea that these molecules are of extraterrestrial origin (Chap. 3).

Developments in radio and infrared astronomy have shown that chemistry is alive throughout the Universe. Old stars, in addition to synthesizing chemical elements by nuclear reactions, are also active chemical factories in making molecules. While nuclear reactions occur in the centers of stars, the chemical syntheses take place in the outer layers of stars. The separate zones represent the different needs for the reactions. Whereas nuclear reactions require high temperatures and densities, molecules can only survive under low temperatures. The chemical elements, in particular carbon, produced in the inner regions of stars are brought up to the surface by convective motions. Once they are in the cool stellar atmosphere, they can react and make simple molecules such as CO. However, the more complex molecules that we see in these old stars are not made in the stars themselves, but in the stellar winds of the stars. The densities under which these molecules are made are less than one-trillionth of that of the Earth's atmosphere, lower than the best vacuum that scientists can create in the laboratory. Because of the high speeds of these winds, stars are constantly dumping the molecules into interstellar space. The fact that these molecules can be detected in the stellar vicinity means that they are constantly synthesized and replenished on time scales of only hundreds of years.

Not only are molecules being made, they are made extremely fast and efficiently. These are incomprehensible feats based on today's theoretical understanding of chemistry. From our experience on studying chemical reactions in the terrestrial laboratory, reactions will only occur under sufficiently high densities (allowing atoms to collide with each other frequently) and under relatively high temperatures (when atoms are moving sufficiently fast). Yet in the environment of the stellar winds, both the density and temperature are low. How these stars manage to perform such chemical miracles has remained a mystery.

Since molecules are the building blocks of solids, can these molecules found in the stellar winds of old stars group together themselves and condense into solids? We know that water vapor in cold Earth atmosphere can condense into snow. Can a similar process work in stars? When the molecules flow away from the star in the stellar wind, temperature in the environment will drop. Will this cooling lead to the formation of solids? Is the presence of silicon and oxygen-based molecules related to the creation of silicate grains that we discussed in Chap. 6? Most significantly, can the many carbon-rich molecules found in carbon stars form organic solids? Answers to this question will be found in the next chapter.

A brief summary of this chapter
Stars make not just elements, but also molecules. These molecules could be the basis for stardust.

Key words and concepts in this chapter
- What is in the air?
- Search for molecules in space
- Rotation and vibration of molecules
- High-frequency radio receivers and millimeter-wave spectroscopy
- Spectroscopic identification of molecules in space
- Old stars as molecular factories
- Discovery of fullerenes in planetary nebulae

Questions to think about
1. The need to detect molecules in space led to the development of high frequency radio receivers reaching frequencies higher than 100 GHz in 1970. The range of frequencies used by television broadcasts (VHF and UHF bands) at the time was lower than 1 GHz. Astronomical radio technologies were hundreds of times in frequency ahead of commercial technology. What do you think of the relationship between astronomy and technology? Do you think that these high-frequency receivers will be deployed in commercial products in the future? What are the practical advantages of using high-frequency technology?
2. Are you surprised that fullerenes, a molecule created artificially in the laboratory, were detected in space? If they can be made in space, why is this molecule not naturally made on Earth?
3. Astronomers only discovered that chemical elements are made in stars in the 1950s, and that molecules can be manufactured in stars in the 1970s. What surprises could be in store for us in the future?
4. Avogadro was one of many scientists whose work was not recognized in their lifetimes. The reasons are many, including the lack of self promotion, not mixing with those who are politically powerful, or simply just being ahead of their time. In the case of Avogadro, there was also the case of geographical isolation. With communications being much more efficient today, do you think there are scientists or ideas today which are suffering from the same fate?
5. The identification of interstellar molecules relies on the comparison between spectral lines observed in celestial objects to molecular lines measured in the laboratory. However, there have been cases where the molecules were identified strictly from astronomical observations before they were confirmed by subsequent laboratory measurements. There are still thousands of spectral lines in astronomical spectra that remain unidentified, although their molecular origin is suspected. What do you think these molecules could be? Will they ever be identified?

Chapter 10
Smoke from Stellar Chimneys

The importance of the fireplace in the American home can be traced back to pioneer days. The fireplace is where the entire family gathered to share warmth and conversations during the long, dark winters. To this day, when modern central heating has all but eliminated the need for a fireplace, our emotional ties ensure that fireplaces remain a permanent fixture in many homes built in North America.

The smoke that escapes from the chimneys of farmhouses is also a familiar sight on prairie landscapes. The gathering and chopping of firewood is such a strong part of our heritage that urban dwellers sometimes spend their weekends in country cottages just to relive the experience.

But how often have we thought about what makes up the smoke generated from the crackling fire? Wood is made of cellulose, the cell walls of plants. The process of burning breaks up these complex organic compounds into simpler, smaller gases, such as water vapor and carbon dioxide. However, firewood is rarely totally dry, and the confined space of the fireplace usually does not supply enough air to burn the wood thoroughly. The result of such incomplete combustion is smoke. When the carbon atoms in wood cannot react completely with oxygen to form carbon dioxide, they tend to congregate into small particles with typical sizes on the order of 1/10,000 of a centimeter. Although the gaseous products of burning are often invisible, these small particles obstruct sunlight, and are seen as black smoke rising from the chimneys.

In terms of chemical makeup, smoke from wood burning is not dissimilar to soot in candle flames or combustion discharges from diesel engines. If we examine the soot particles from flames or combustion under an electron microscope, we find that they are made of chains of spherical units each containing about 100,000 carbon atoms and several tens of thousands of hydrogen atoms. Scientists studying flames suggest that gas molecules produced by burning first aggregate into very small particles, which later collide and coalesce into larger ones.

In the last several chapters, we have learned that old stars produce gases and small solid particles of micrometer sizes. From these observations, it would seem

S. Kwok, *Stardust*, Astronomers' Universe, DOI 10.1007/978-3-642-32802-2_10,
© Springer-Verlag Berlin Heidelberg 2013

that old stars are like giant fireplaces or furnaces, giving off gases and small solid particles through their atmospheres. Are these stellar exhausts the result of breaking down complex matter as in a fireplace, or do they make the smoke through a different process?

Since the early twentieth century, astronomers have suspected there are small solid particles in interstellar space (Chap. 6). After systematic measurements of the colors of stars, astronomers gradually came to accept the possibility that small solid particles existed in the Galaxy. Since small particles deflect more blue light than red light, astronomers found that the abnormal redness of some stars is not due to intrinsic characteristics of the stars but is the result of intervening material between the stars and us. An analogy for the above effect can be found in our everyday experience. The Sun appears redder at sunset because we are looking through a longer path of atmosphere, and the effect of small particles in the air is larger. After a forest fire, the sunset is especially spectacular because of the increased amount of smoke particles in the air. Astronomers were able to quantitatively work out that the sizes of these interstellar solid particles are on the order of 1 μm, or about 100th of the thickness of a strand of hair.

What is the chemical makeup of such small particles? One way we can tell is by seeing whether the absorption of starlight carries any signature corresponding to known materials. Every solid material is made of atoms and the collective motions of these atoms give rise to specific spectroscopic signatures. By comparing the astronomical spectrum and laboratory measurements of minerals, one can identify the chemical nature of interstellar dust. Unfortunately, there was no definitive signature in the visible part of the spectrum that gave us any clue.

In the early 1960s, astronomy entered its space era through the placement of telescopes on rockets. A 32-cm telescope was placed on an Aerobee rocket which carried a gyroscopic stabilizer that allowed the telescope to point accurately at a star. With brief 20-s observations, the telescope scanned through the ultraviolet spectrum of two stars in the constellation of Perseus. The results, as reported by Theodore Stecher of NASA Goddard Space Flight Center in Maryland, showed that these two stars suffered from a sharp drop of light in the ultraviolet at the wavelength of 0.22 μm (or 220 nm), which can be attributed to the effect of absorption by small solid particles between us and the stars. This absorption feature in the ultraviolet therefore became the first clue to the chemical makeup of interstellar dust.

Almost 50 years have passed since the discovery of the 220 nm feature, and there is still no unanimous agreement among astronomers on the chemical nature of the carrier of this feature other than the fact that it is a carbon-based compound. May be it is an exotic organic material that is not present on Earth, or the feature is produced only under the extreme conditions in interstellar space.

Another important question facing astronomers is: Where do these small particles come from? The discovery of stardust, or smoke from stellar chimneys, came as a complete surprise to astronomers. As we related in previous chapters, the development of infrared astronomy has clearly shown that old stars can produce solid particles (smoke) in abundance very quickly. The identification of these

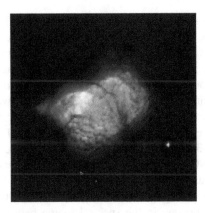

Fig. 10.1 *Hubble Space Telescope* image of NGC 7027.
Planetary nebulae are beautiful nebulous objects that all stars (including our Sun) will become near the end of their lives. The planetary nebula NGC 7027 in the constellation of Cygnus was the first celestial object to be found to have complex organic compounds with aromatic structures. Image credit: Robin Ciardullo

particles as forming from silicates demonstrated that inorganic minerals can be made by stars. In Chap. 5, we talked about the other common stardust silicon carbide, which contains carbon. Since carbon is the fourth most common chemical element in the Universe, and is being produced by the nuclear furnaces inside carbon stars, it was also speculated that graphite, a pure carbon substance, could also be a form of stardust.

All the stardust that has been identified or is under discussion so far—silicates, silicon carbide, and graphite—are all mineral-like inorganic substances. What I am about to describe in the remainder of this chapter is probably among the most surprising astronomical discoveries of the twentieth century. Instead of graphite, infrared spectroscopic observations have found stardust made of pure organics—the organic compounds that are the basic ingredients of life. The discovery was entirely unexpected and its significance not appreciated for a long time.

In the early 1970s after the University of Minnesota developed the infrared detectors and had some success using them on the 0.8-m telescope near Minneapolis (Chap. 6), they needed a larger telescope on a better site. Together with the University of California at San Diego they built a 1.5-m telescope in Mt. Lemmon, near Tucson, Arizona. In 1973, Fred Gillett, W. J. Forrest, and Mike Merrill of the University of California at San Diego (UCSD) observed the planetary nebula NGC 7027 (Fig. 10.1). Planetary nebulae are colorful objects formed from the ejecta of old stars and NGC 7027 is known to descend from carbon stars such as CW Leo (Chap. 8). In addition to the atomic lines expected, they were surprised to find a strong feature in emission at the wavelength of 11.3 μm. In November 1974, Mike Merrill and Ray Russell, both graduate students at UCSD, wanted to test the newly built near infrared spectrometer and observed NGC 7027 from Mt. Lemmon. They had a strip chart recorder connected to the data system, and as the filter wheel stepped through the wavelength range while observing NGC 7027, suddenly the pen started going off scale near the wavelength of 3 µm. They were concerned that there was something wrong with the instrument, so they stopped and checked to make sure it had not run out of coolant, which it had not. They checked cables and power supplies and found nothing wrong. They then tried taking the spectrum again, and saw the same thing happening. They then lowered the gain of the amplifier to make sure that the measurement would not go off scale. A standard star was observed and everything was found to be normal. They therefore came to the conclusion that the 3 µm feature was not an instrumental effect and must be real, intrinsic in the object NGC 7027. This was the first discovery of the 3.3 µm feature.

Since two mysterious features have been found at 3.3 and 11.3 µm, it would be useful to obtain a more complete infrared spectrum of NGC 7027. However, the Earth's atmosphere is opaque in the spectral region between 5 and 10 µm, and observations cannot be taken from a telescope on the ground. In May and November of 1976, the UCSD team had a chance to take their spectrometer to NASA's *Kuiper Airborne Observatory,* a flying observatory converted from a C-140 military transport plane (Fig. 7.3). They first saw the expected atomic lines of magnesium, so they knew the system was working. Then they discovered new features at 6.3 and 7.7 µm. Together with the 3.3 and 11.3 µm features, they were named the unidentified infrared (UIR) features.

The observation of the UIR features shows that when we open a new spectral window to the Universe, there can be unexpected serendipitous discoveries. The 1970s represent the first opening of the infrared spectral window for spectroscopic observations. This opening confirmed some known knowledge—for example, the emission lines from atoms. Unlike atomic lines, the UIR features are broad (Fig. 10.2). They are also bright, implying that whatever material that is emitting these features, there must be a lot of it. This in turn suggests that the emitting substances must be made of common ingredients.

These strong emission features represented a great puzzle for astronomers. The widths of the features are very broad, suggesting that they do not originate from atoms or molecules. The wavelengths at which the UIR features occur, at 3.3, 6.2, 7.7, and 11.3 µm also do not correspond to any substances known to astronomers. But in fact these features have indeed been seen in the laboratory and they arise from organic compounds. However, most astronomers were not familiar with the literature of organic chemistry and it had not occurred to them that an organic substance could be present in interstellar space.

The first suggestion that these features could arise from organic compounds was made by Roger F. Knacke of the State University of New York in Stony Brook in 1977, but his paper in *Nature* was completely ignored by the astronomical community. In 1979, the UCSD team consisting of Rick Puetter, Ray Russell, Tom Soifer, and Steve Willner published a paper in the *Astrophysical Journal* suggesting that stretching and bending modes of hydrocarbon compounds may be responsible. Walt Duley, a professor of physics at York University in Canada (now working at the University of Waterloo in Canada) had done contract work with General Motors on automobile exhaust and was familiar with the properties of soot and other chemicals

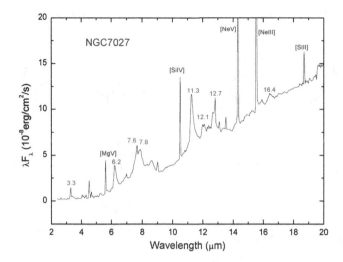

Fig. 10.2 The UIR features in NGC 7027.
Infrared spectrum of the planetary nebula NGC 7027 taken by the *Infrared Space Observatory*. The vertical scale is brightness and the horizontal scale is wavelengths in units of micrometers (μm). The unidentified infrared (UIR) features are labeled by their wavelengths (e.g., 3.3, 6.2, 11.3). The narrow, sharp lines are atomic lines due to magnesium, silicon, neon and sulfur. Although the UIR features are as bright as some of the atomic lines, their shapes are much broader than the atomic lines

coming out of exhaust pipes of cars. In 1981, he and David Williams of the University of Manchester, England (now at University College London), made the astonishing suggestion that the UIR features originate from aromatic compounds. Aromatic compounds are molecules with ring-like structures; the most common everyday examples are benzene and toluene, which are derived from coal tar. Benzene (C_6H_6) is made of six carbon atoms in the shape of a ring, with each carbon atom having a hydrogen atom attached. Toluene ($C_6H_5CH_3$), a widely used industrial solvent, is similar to benzene except that one of the hydrogen atoms is replaced by a CH_3 group. Aromatic structures are paramount in biomolecules, which are the molecules responsible for the biological functions of life. DNA and RNA bases (adenine, thymine, cytosine, guanine, and uracil) are all built upon the basic aromatic molecules purines or pyrimidines. Proteins are built from amino acids and some of the amino acids, such as histidine, phenylalanine, tryptophan, and tyrosine, also have aromatic units as their building blocks. Purines, pyrimidines, and amino acids are all examples of prebiotic molecules.

Louis Allamandola, a chemist working at the NASA Ames Research Center in California, who previously worked at the University of Leiden in the Netherlands and became familiar with the problem of UIR features, came to a similar conclusion several years later. He suggested the UIR features were specifically due to polycyclic aromatic hydrocarbon (PAH for short) molecules, the simplest of the aromatic compounds family (Fig. 10.3). This idea was echoed across the Atlantic. Jean-Loup

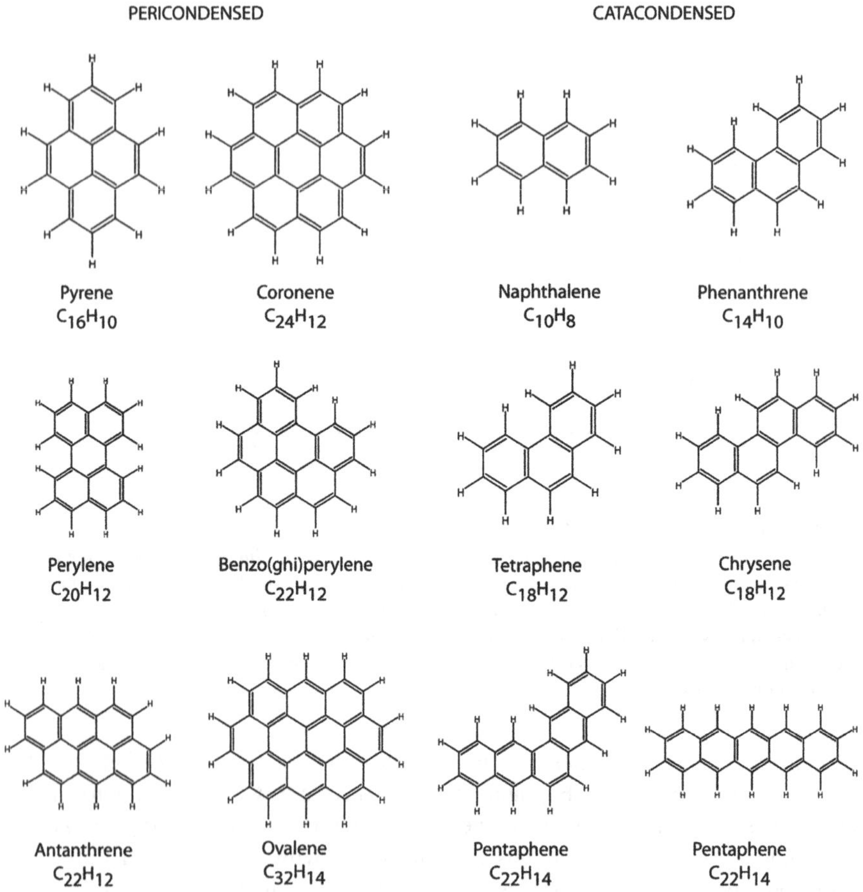

PERICONDENSED   CATACONDENSED

Pyrene   Coronene   Naphthalene   Phenanthrene
$C_{16}H_{10}$ $C_{24}H_{12}$ $C_{10}H_8$ $C_{14}H_{10}$

Perylene   Benzo(ghi)perylene   Tetraphene   Chrysene
$C_{20}H_{12}$ $C_{22}H_{12}$ $C_{18}H_{12}$ $C_{18}H_{12}$

Antanthrene   Ovalene   Pentaphene   Pentaphene
$C_{22}H_{12}$ $C_{32}H_{14}$ $C_{22}H_{14}$ $C_{22}H_{14}$

Fig. 10.3 Examples of PAH.
Examples of some simple polycyclic aromatic hydrocarbon molecules (naphthalene, anthracene, pyrene, chrysene, perylene, benzoperylene, coronene, ovalene). PAH molecules are made of pure hydrogen and carbon atoms and consist entirely of rings. These rings can be arranged in a variety of geometric ways as seen in the examples above

Puget of Ecole Normale Superiéure and Alain Léger of University of Paris VII in France believed that PAH molecules containing less than 50 carbon atoms could account for the infrared emissions observed in the diffuse interstellar medium.

Astronomers were initially skeptical of the aromatic compound identification proposed by Duley and Williams because most found it hard to believe that organic compounds of such complexity could be present in space. In the PAH idea, they saw a way out. After all, molecules were then known to be common in interstellar space (Chap. 9). PAHs are molecules, just with a few atoms more than the largest molecules known at the time. The PAH hypothesis soon gathered a following and developed into a bandwagon. Hundreds of scientific papers were written on the subject and the term PAH became a popular catch phrase in the astronomical literature.

However, the community was not without its doubters. One of the early voices questioning the PAH idea was Bert Donn of the Goddard Space Flight Center in Maryland. At the International Astronomical Union symposium on interstellar dust held in Santa Clara, California in July 1988, Donn raised a number of questions and expressed his opinion that the PAH hypothesis lacked the supporting evidence to warrant such widespread acceptance and dominance as the cause of the UIR phenomenon. Unfortunately Donn's criticism never made it into the astronomical journals.

While astronomers were enthusiastic about PAH as the solution to the UIR mystery, chemists were less so. PAHs are simple molecules and their spectral properties have been well studied both theoretically and experimentally. Richard Saykally and his group at the University of California at Berkeley designed an experiment to measure the emission spectra of PAH molecules under simulated space conditions. While there are qualitative similarities between PAH spectral features and astronomical UIR features, they do not match in detail. In a paper in the *Astrophysical Journal* in 2000, they concluded that "no PAH emission spectrum has been able to reproduce the UIR spectrum with respect to either band positions or relative intensities". The Berkeley group suggested that if PAHs are responsible for the UIR features, they cannot be in a neutral form and an ionized state of PAH offers a better chance of matching.

In spite of the extreme popularity of the PAH hypothesis, I find it lacking in specifics. While there is good evidence that the UIR features arise from aromatic materials, they are not necessarily PAH molecules. Because PAH are molecules, their vibrational lines are sharp and well defined. These do not resemble at all the astronomical observations as the observed UIR features are broad (Fig. 10.2). To explain the astronomical observations, the PAH believers had to appeal to a large mixture of PAH molecules with different chemical compositions and sizes, and the fitting results seemed artificial. Many attempts were made to search for the individual PAH molecules in the interstellar medium. In spite of their well-known frequencies, none was detected.

Since the UIR features are not seen in carbon stars, their detection at a later stage of evolution—in planetary nebulae—implies that the compounds that are responsible for emitting the UIR features are made in the circumstellar envelopes of the stars. From the theory of stellar evolution, we know that only a few thousand years separate the stages of carbon star and planetary nebula. Hence, the aromatic compounds must have been transformed from simpler molecular forms over a similar timescale.

In order to further pinpoint the chemical process, it would be useful to study the transition objects between carbon stars and planetary nebulae. These objects, called proto-planetary nebulae (Fig. 10.4), are in a state of rapid evolution and live only a thousand years. The *ISO* satellite and its capabilities are well suited for these studies. By observing proto-planetary nebulae that we had previously discovered (Chap. 8), Kevin Volk, Bruce Hrivnak, and I found that they have a much richer infrared spectrum. In addition to the aromatic features, we also detected signatures of aliphatic chains attached to the aromatic rings. Aliphatic chains are

Fig. 10.4 *Hubble Space Telescope* image of the Water Lily Nebula. The Water Lily Nebula is one of many proto-planetary nebulae discovered and imaged with the *Hubble Space Telescope* by Sun Kwok and Bruce Hrivnak. Together with Kevin Volk, they used the *Infrared Space Observatory* to find that these nebulae are extremely proficient in producing complex organic matter

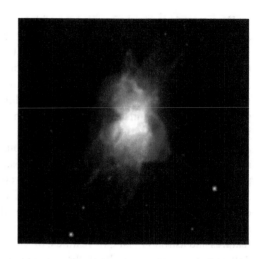

hydrocarbons with a linear structure, which are very different from the ring structures of aromatics. What excited me the most is that aliphatics form the basis of biomolecules such as fats and oils. Animal fats (e.g., butter, a solid) and vegetable oils (e.g., corn and peanut oil, liquids) have closely related chemical structures and together they form the general class of biomolecules called lipids. Our detection of aliphatics, in addition to the known aromatics, means the stars are capable of manufacturing the basic ingredients of life.

Together with Peter Bernath of the University of Waterloo in Canada (now at Old Dominion University in Virginia), we theorized that when these substances formed, they were made in a disorganized manner with a mixture of chemical structures. First formed are small groups of aromatic rings and a variety of aliphatic chains got attached to these rings. With time, under the influence of ultraviolet light from the star, they lose part of their outer structures and the substance becomes aromatic.

At the same time, the team at Ohio State University led by Kris Sellgren found that the UIR features in three reflection nebulae NGC 1333, NGC 2068 (Fig. 10.5), and vdB 133 have virtually identical shapes and relative strengths, although the central stars of these three nebulae have temperatures ranging from 6,800 K for vdB 133, to 11,000 K for NGC 1333, to 19,000 K for NGC 2068. Since only hot stars (those with temperatures exceeding 30,000 K) emit a significant amount of ultraviolet radiation, the central stars of all these three reflection nebulae do not feed a significant amount of ultraviolet to the surrounding nebulae.

Sellgren began her study of reflection nebulae when she was a graduate student at Caltech. During her career as professor at the University of Hawaii and Ohio State University, she developed into a leading expert in the study of reflection nebulae. In 2011, she made news headlines for the discovery of fullerene (C_{60}) in the reflection nebula NGC 7023 (Chap. 9). In addition to her career as a research astronomer and teacher, Sellgren also has an active career as a musician, playing the guitar and composing a number of songs. Among the songs she wrote are "Ala Moana", "Android Blues", and "California Lost".

Fig. 10.5 The reflection
nebula NGC 2068.
The reflection nebula NGC
2068 (Messier 78) in the
constellation of Orion.
Reflection nebulae (the blue
color object in the picture)
shine by the scattering of star
light by solid particles in the
nebulae. Image credit:
European Southern
Observatory

The detection of the UIR features in proto-planetary nebulae is significant because these objects have low temperature central stars and they hardly emit any ultraviolet light. Yet one of the basic premises of the PAH model is that these molecules are excited by ultraviolet light. In order to get around this problem, the PAH supporters had to revise the model to include molecules in positive and negative charge states, or include large clusters of molecules. These modifications would make the PAH molecules more receptive to visible light and therefore more in line with the observed reality.

A group of scientists led by Renaud Papoular at the Centre d'Etudes de Saclay in France, took great interest in the astronomical mystery of the UIR bands. Coal is abundant in France and has been extensively mined for fuel until the 1980s. The Centre de Recherche sur le Charbon studied the various stages of evolution of coal from living organisms to graphite. Papoular compared the infrared spectrum of coal in the book "*Les Carbones*", written by a group of French physicists and chemists called "Groupe Français d'Etude des Carbones", and found that the coal spectrum bears a great resemblance to the astronomical UIR spectrum. Papoular's group compared our infrared observations of proto-planetary nebulae with those of coal and found amazing similarities. Coal (Fig. 10.6) is composed of a mixture of aromatic rings and aliphatic chains. Its black color also correlates well with the astronomical observations.

In the meantime, Françoise Behar and M. Vandenbroucke of the Institut Français du Pétrole in Rueil-Malmaison, France, provided samples of kerogen to Papoular's group and the Saclay team began to consider kerogen as another alternative carrier of the UIR features. Kerogen is an insoluble, tar-like, organic compound distributed in rocks (Fig. 10.7). In terms of chemical composition, kerogen is a mixture of complex macromolecules which are the remnants of ancient life (Fig. 10.8). After burials in sediments, kerogen underwent gradual evolution, breaking up into lighter molecules as well as condensed residues. The evolution of kerogen is believed to be the source of fluid hydrocarbons such as oil.

Fig. 10.6 A picture of coal.
Coal is extensively mined all over the world and is commonly used for fuel. Its chemical composition is extremely complex, with aromatic and aliphatic units connected in random patterns. Coal is believed to have originated from remnants of life long ago

Fig. 10.7 A picture of kerogen embedded in rock.
Kerogen is a natural organic compound containing hydrocarbons and impurities such as oxygen, nitrogen and sulfur. Like oil and coal, it has the potential to be used as fuel except that it is difficult to extract. Solid kerogen, when heated, can be transformed into liquid shale oil. After the removal of sulfur, nitrogen, and other impurities, shale oil can be used the same way as crude oil. Photo credit: Colorado Geological Survey

These suggestions at first seemed totally incredible. Coal is supposed to be made from remains of living things under high pressure inside the Earth, far different from the low-density environment in space. Astronomers also found it aesthetically unappealing that a dirty, black substance could be proliferating among heavenly bodies. In a way, the community is still hung up on the Aristotelian idea that the heavens should be a purer, simpler, and cleaner place than Earth. For years, Papoular preached at various conferences to skeptical audiences who refused to consider the possibility of coal-like substances in space.

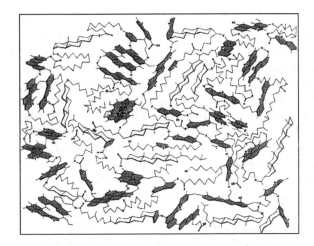

Fig. 10.8 A schematic drawing of the chemical structure of kerogen.
The dark patches are aromatic rings and the lines are aliphatic chains. Each corner of the rings and the chains has a carbon atom. The hydrogen atoms are attached to the carbon atoms but are not shown explicitly. We can see that the aromatic rings are in the form of islands of several rings which are connected to each other by aliphatic chains. The orientations of the rings are random and the entire structure is disorganized. Chemists call such structures amorphous. Image credit: Papoular (2001)

However, the spectral resemblance between laboratory and astronomical measurements is so strong that even if the substance is not a type of coal, it is probably something similar. For example, Walt Duley has used lasers to blast graphite and create a synthetic material called hydrogenated amorphous carbon (HAC) which also shows some of the observed features. HAC does not contain impurities such as oxygen and nitrogen as coal does, but the chemical structures of the two are not that different.

In terms of structure, coal and HAC have one thing in common. They are both considered amorphous materials. Amorphous is a term used by chemists to refer to materials that have disorganized internal structures. Most solids are in the crystalline form where the atoms are arranged in periodic order and have precise spatial relationships with each other. A simple example is salt, which contains sodium and chlorine atoms arranged in a cubic structure of fixed sizes and angles. An amorphous solid, on the other hand, contains atoms that are arranged randomly with no clear pattern or order. Not only the spacing, but the orientations of atoms with respect to each other are also random. A familiar example is glass. Glass is made up of a network of silicon and oxygen atoms but its chemical structure is irregular.

Coal is a natural substance and HAC is something artificially synthesized in the laboratory. Another artificial amorphous substance, called quenched carbonaceous composites (QCC), was made in the laboratory by Akira Sakata of Japan in 1984. QCC is a dark, granular material that results from heating methane gas to 3,000 K in a microwave oven and allowing it to cool and condense at room temperature. At first, Sakata was motivated to find a new substance to explain the 220 nm ultraviolet absorption feature seen in the interstellar medium. However, he found that QCC has interesting infrared properties and his team performed a series of experiments to refine their understanding of what QCC was. In 1985, Sakata attended a workshop on interstellar dust in Hilo, Hawaii. After the meeting, he went to see Alan Tokunaga, an American astronomer of Japanese descent, at the University Hawaii in Honolulu and told him about his new material QCC as a possible component of interstellar dust and that was the beginning of a fruitful collaboration. Tokunaga performed astronomical observations with a new infrared spectrometer at NASA's 3 m *Infrared Telescope Facility* at Mauna Kea, Hawaii, to compare with the laboratory spectra obtained from QCC by Sakata.

Sakata-san graduated from Tokyo Metropolitan University as a chemistry student studying geology. In his early days, he studied the chemical composition of water from hot springs. Unlike typical Japanese academics, he spoke loudly and directly and did not conform to the polite conventions of Japan. Fortunately for astronomy, he did not get along with his professor and later switched to interstellar chemistry. At the University of Electro-Communications in Tokyo, he set up his laboratory apparatus with Setsuko Wada which led to the discovery of QCC. Sadly their collaboration was cut short by Sakata's untimely death from stomach cancer in 1996.

The way that QCCs are made is very simple. Sakata started with a plasma tube containing low-pressure methane which was excited to high temperatures by a microwave generator. The gas was allowed to expand into a vacuum chamber through a small nozzle. Condensing on a room-temperature substrate as well as on the walls of the vacuum chamber was the new, dark, carbonaceous material called QCC. Sakata and his team were able to create several different varieties of QCC. One is a brown-black material which he called "dark QCC", another is a yellow-brown filmy material called "filmy QCC" which condenses on the wall. In the center of dark QCC is a black circular spot about a few nanometers wide which Sakata gave the name "granular QCC".

After the death of Sakata, his work was carried on by Setsuko Wada, who continued to collaborate with Tokunaga. With modern electron microscopes, the shape of these novel materials can be seen and some pictures are shown in Fig. 10.9. By bombarding QCC with X-rays and observing the electrons released as a result, Wada was able to determine the chemical structure of the material and found that QCC consisted of networks of small aromatic rings and aliphatic chain-like side groups, similar to coal. Wada also found that QCC is made up of more than hydrogen and carbon, but contains a dose of oxygen as well. In this respect, QCC is also similar to coal.

It is interesting to note that the laboratory procedure Sakata used to make QCC is not unlike the physical conditions in the stellar winds of old stars. There, hydrocarbon molecules are made and ejected from the stellar atmosphere under high temperature. The gas cools as the wind is expanding at supersonic speed. If there are other small solids that can serve in the role of substrates in the laboratory, it is likely that amorphous compounds like QCC can condense onto these solid surfaces.

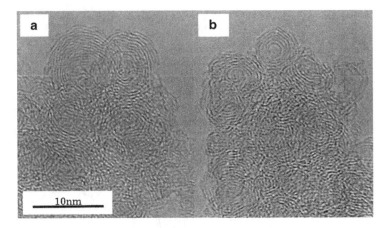

Fig. 10.9 Electron microscope pictures of QCC.
The image on the left is granular QCC and the one on the right is dark QCC. We can see that this artificial substance is made up of groups of onion-like particles. The scale on the lower left shows the size of 10 nm. Photo credit: Setsuko Wada

When the properties of QCC are studied in the laboratory, they show a lot of similarities with the spectral properties of soot. Since soot is a natural product of burning and is not produced under controlled laboratory conditions, the term "soot" really encompasses many different sizes and shapes of carbonaceous materials. These different morphologies and structures were studied by the teams of Vito Mennella in Osservatorio Astronomico di Capodimonte in Napoli, Italy, and Louis d'Hendecourt in Institut d'Astrophysique Spatiale, in Orsay, France. No matter what the chemical makeup of stardust is, it shares some common properties with the products of combustion on Earth.

Independent of the above efforts, another group in Italy led by Franco Cataldo experimented with petroleum fractions as a possible carrier of the UIR features. These fractions are extracted from petroleum during the refinement process. By subjecting several different kinds of petroleum extracts to analysis, Cataldo found that their infrared spectra resemble those seen in proto-planetary nebulae. This is not totally surprising as we know that both coal and oil have common ancestors and their chemical structures are similar.

While the spectral match seems convincing, it is difficult to accept petroleum being responsible for the UIR features as petroleum originates from the decomposition of formerly living matter. The transformation to petroleum is the result of biochemical processes involving bacteria and fungi, as well as physical processes such as intense temperature and pressure. These processes took millions of years. How can stars manage to produce petroleum-like matter so quickly?

Except for these few dissenting voices, the PAH hypothesis became extremely popular in the astronomical community. Although work on alternate models of the UIR phenomenon continued to be published, these results were ignored. In May 2010, a conference was organized in Toulouse, France, to celebrate the 25th

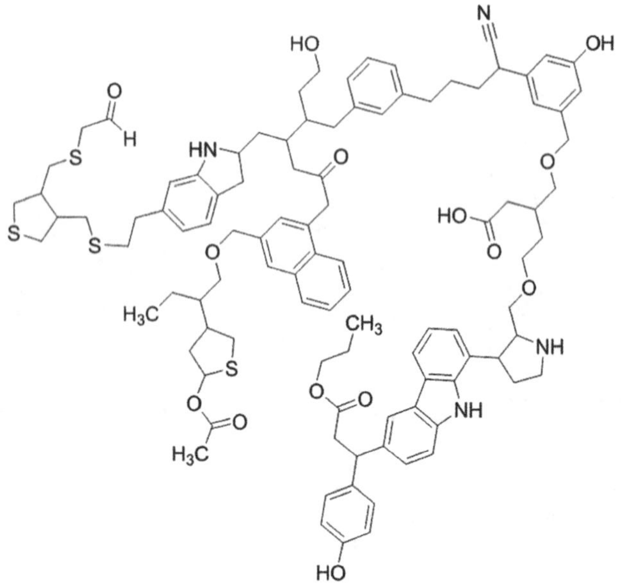

Fig. 10.10 Schematic structure of MAON.
This typical structure is a mixture of rings (aromatic) and chain (aliphatic) chemical sites. Each corner junction of the lines is a carbon atom. Hydrogen atoms are not shown. There are about 100 carbon atoms in this schematic. An actual particle would consist of multiple units of similar structures

anniversary of the introduction of the PAH hypothesis. While most of the talks sang the praises of the PAH idea, the conference did not go without dissenting voices. Comments that other carbonaceous materials could be present were raised by Emmanuel Dartois of the Institut d'Astrophysique Spatiale, University of Paris-South in Orsay, France. The role of the aliphatic component has to be taken into consideration as a component of the chemical structure of the emitter of the UIR features.

The publication of our paper "Mixed aromatic/aliphatic organic nanoparticles as carriers of unidentified infrared emission features" in the journal *Nature* in 2011 was the first official challenge to the PAH hypothesis. Using data from proto-planetary nebulae and novae, we showed that the aliphatic component is indeed the dominant chemical component of the UIR features. The structure of this carrier must be disorganized, not neat and pure PAH molecules. When these complex organics form in the winds of old stars, we argued, the products are likely to be untidy and all kinds of junk will hang on the sides of the aromatic rings. These complex organics, which we call "MAON" (standing for "mixed aromatic/aliphatic organic nanoparticles") for short, will also likely have incorporated other common elements such as oxygen, nitrogen, and sulfur. A schematic unit structure typical of MAON is shown in Fig. 10.10. Such natural products give a much better match to the observed astronomical spectra of the UIR bands.

This debate on the nature of the carrier of the UIR bands will likely go on for some time. Do the features arise from a collection of free-flying PAH molecules, or do they come from a more complicated chemical system such as MAON? Whatever the exact chemical structure of the carrier of UIR is, there is no doubt that complex organics are present throughout the Universe. How these complex organics came about and what the implications of the presence of these organics are will be addressed in the later chapters.

A brief summary of this chapter
Stardust also contains organic compounds with complexity similar to coal and kerogen. The exact form of the chemical structure of these organic compounds is still under debate.

Key words and concepts in this chapter
- Stars are like fireplaces that give out smoke and soot
- The 220 nm unidentified absorption feature in interstellar dust
- Unidentified infrared (UIR) emission features discovered in planetary nebulae
- UIR features due to polycyclic aromatic hydrocarbon molecules?
- Aromatic and aliphatic compounds found in proto-planetary nebulae
- UIR features seen in natural substances such as coal, kerogen, and petroleum fractions
- UIR features also seen in artificial substances such as hydrogenated amorphous carbon and quenched carbonaceous composites
- Debate continues on whether the UIR features are due to PAH molecules or disorganized, complex organics as in the MAON model.

Questions to think about
1. Soot and smoke are products of burning. Does the fact that planetary nebulae produce substances similar to soot and smoke suggest that there are also things burning in planetary nebulae? Is the physical mechanism of producing soot and smoke similar or different in fire places and planetary nebulae?
2. Coal and oil are remnants of ancient life (Chap. 14), but the chemical structures of artificial substances such as HAC and QCC bear some resemblance to coal and oil. Is this a coincidence?

Chapter 11
Gems from Heaven

Precious stones and the gems that are cut from them have been used throughout human history as signs of wealth, luxury and power. They are distinguished from other minerals found on Earth by their color, transparency, and refractive qualities. Their physical hardness and resistance to environmental elements give them a sense of ever-lasting value. While these properties give gems the qualities as ornamental objects, it is, above all, rarity that commands the value assigned to them as objects of treasure. Although pebbles on the beach also come with a variety of colors and can be considered quite beautiful, they are not valued as such just because there is such an abundance of them.

Myths and legends associated with gems have evolved in many different cultures. Many are thought to have magical powers, thus elevating the wearer to special status. Some are thought to have medicinal qualities that can cure or protect the wearer from diseases. Most interestingly, gems are often thought to be of celestial origin. The progress in science in the last 200 years has all but removed these fantastic ideas from our thinking. The development of geology has shown the precious stones to be nothing more than minerals that are part of the Earth's crust. Modern chemical studies have revealed that there are no special substances in gems: they are made up of the same chemical elements, such as carbon, oxygen, silicon, aluminum, magnesium and iron, as common rocks. Even their luster and colors can be understood through physical principles such as refraction, reflection, and dispersion. From a chemical point of view, gems are no different than ordinary pebbles, which are made up primarily of silicon dioxide, mixed with some iron oxide and small amounts of manganese, copper, aluminum, magnesium, and other impurities. The internal structures of gems are now accurately determined through the technique of X-ray crystallography. We now know that, with a few exceptions such as amber and opal, precious stones are crystals made up of groups of atoms in a periodic structure. They are arranged in various geometric patterns, giving them their unique optical properties.

The amazing thing is that we may have come a full circle regarding the origin of gems. There are certainly good theories on the genesis of most minerals on Earth. For example, diamonds are believed to be made at great depths in the Earth's crust

S. Kwok, *Stardust*, Astronomers' Universe, DOI 10.1007/978-3-642-32802-2_11,
© Springer-Verlag Berlin Heidelberg 2013

under extremely high temperature and pressure, and are brought to near the surface by volcanic activities. Rubies and sapphires are formed by reactions between molten aluminum rock and limestone. While there is no doubt that such processes do occur, scientists are now pondering the possibility that some microscopic forms of these gems might have come from space.

The first evidence for the existence of gems and other minerals of celestial origin is from the study of meteorites. In 1987, diamonds of nanometer (10^{-9} m) size were identified in meteorites. This was followed by the discoveries of silicon carbide (SiC) and graphite. The isotopic content of these carbon-based materials suggests that they did not arise from the Solar System, but probably from stars. In 1997, a large amount of corundum (a class of mineral that both rubies and sapphires belong to) was found in meteorites by scientists at the Washington University at St. Louis. Again, these stones are found to be stellar in origin (Chap. 16).

If these grains indeed came from outside the Solar System, where were they made and how did they get here? From the study of the spectra of old stars, astronomers have been able to establish a definite link between minerals found in meteorites and the ejecta of old stars (Chap. 6). We have learned that certain old stars can manufacture vast quantities of solid minerals and distribute them throughout the Galaxy. The discovery of this link between Heaven and Earth is one of the most remarkable stories of modern astronomy.

How do stars actually make these minerals? We know that matter changes from gas to liquid to solid as temperature decreases and density increases. The most familiar example in our everyday life is the change from steam to water to ice. However, the atmospheres of stars are hot (thousands of degrees) and their densities are far below the atmospheric densities on Earth. Both conditions do not favor the formation of solids. The first concern was addressed by Robert Gilman of the University of Minnesota. He argued that due to the low density, the relevant temperature is not so much about the temperature of gases running around but the amount of radiation that materials receive and emit. In our everyday life, our feeling for temperature is based on how often the air molecules hit our body. But if we moved to an airless environment such as the surface of the Moon, our feeling of hot and cold would be strictly governed by whether the Sun is up or not. The same applies to the stellar atmosphere. A mineral may not be able to condense from the gas phase right on the surface of a star, but if it is located some distance from the star, it receives less radiation and cools rapidly. Since silicates and silicon carbides absorb visible light poorly but are strong emitter of infrared light, these minerals are heated less efficiently but cooled effectively. Based on his calculations, Gilman concluded in 1974 that silicates and silicon carbides should be the first minerals to condense in the atmospheres of red giant stars.

If solid particles of silicate and silicon carbide can condense in stellar atmospheres, could other minerals be present too? From the theory of thermodynamics, the first solid materials expected to condense are those with high melting points. Particular interests are paid to aluminum and titanium oxides not only because of their high melting points, but also because of the relatively high abundance of these elements. From an analysis of the *IRAS LRS* data, Irene Marenin of Wellesley College detected a feature at 13 μm in the spectra of a number of red

Fig. 11.1 Pictures of rubies (*left*) and sapphires (*right*).
Although rubies and diamonds are considered to be gems, from a scientific point of view they are nothing but minerals. Corundum (Al_2O_3) is the hardest natural substance after diamond. It is colorless when pure, but when mixed with chromium, it will take on a reddish appearance and is known as ruby. When mixed with iron or titanium, it appears blue and is known as sapphire. Photo credit: David Eicher

giants. A group of Dutch astronomers from the University of Amsterdam suggested that it could be due to an amorphous form of aluminum oxide. The crystal form of aluminum oxide is known as corundum, which is a colorless mineral naturally present on Earth and is the hardest known natural substance after diamond. This idea was followed up by Bill Glaccum of the University of Chicago, who from observations made with NASA's *Kuiper Airborne Observatory*, suggested that this mysterious feature at 13 μm is actually sapphire. Sapphire (Fig. 11.1) is a form of corundum where some of the aluminum atoms are replaced by titanium atoms. As a result, the crystal takes on a bluish color.

These suggestions, although interesting, were difficult to prove due to a lack of good laboratory measurements of corundum and sapphire in the infrared. There are not many laboratories in the world that can do these measurements, and one that can is the Max-Planck Institute in Jena, Germany. Scientists at Jena considered rutile (titanium dioxide, TiO_2), and spinel (magnesium aluminum oxide, $MgAl_2O_4$), in addition to corundum, as the possible carrier of the 13 μm feature. Rutile is not an unreasonable possibility because the molecule titanium oxide (TiO) has been known to be widely present in the optical spectra of old stars rich in oxygen. On Earth, sometimes very fine needles of rutile are found in rubies and sapphires, giving them a star-like appearance.

After the launch of the *ISO* satellite, the Jena group, working with Austrian scientists from Vienna, compared their laboratory results with the superior quality data from *ISO* and concluded that the most likely candidate for the 13 μm feature is spinel ($MgAl_2O_4$). Spinels (Fig. 11.2) on Earth can come in a variety of colors, from red to blue, as a result of some of the magnesium atoms being replaced by iron, zinc, or manganese, and the aluminum atoms being replaced by iron and chromium. Red spinels are sometimes confused with rubies. For example, the "Black Prince's Ruby" in the English crown jewels is in fact a spinel, not a ruby.

Although we have probably not heard the last of the debate on this issue, there is however no doubt that minerals are involved. Whether they are sapphires, rubies or spinels, old stars have clearly demonstrated their abilitiy to synthesize them without difficulty, even under most unforgiving (high temperature and low density) conditions.

Fig. 11.2 Crystals of spinel. Like corundum, pure spinels are colorless. The colors of red and pink in spinel are due to chromium, green is due to iron, and blue is due to zinc. Photo credit: David Eicher

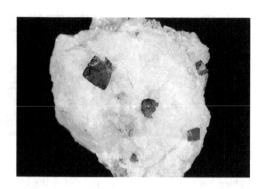

In a completely separate development, laboratory examinations of interplanetary dust particles have revealed a population of sub-micrometer sized glassy silicate grains. The scientists who discovered them called this material GEMS, standing for "glass with embedded metal and sulfides". The name GEMS created quite a stir in the press but strictly speaking this material is more similar to the silicate grains seen in the interstellar medium (Chap. 6) than the gemstones as we know them in our everyday life. Laboratory infrared spectroscopy of GEMS shows that the material shows the 10 µm feature similar to that seen in interstellar silicate grains. Although glass, which is made up of silicon and oxygen atoms, can be quite pleasant to look at, it does not qualify as a gem since glass is commonly available on Earth. This, however, illustrates that the creative use of acronyms can go a long way in promoting one's results!

So far we have emphasized the use of infrared techniques to study stellar minerals. But are there other ways that minerals in space can be identified? What infrared observations offer is that we can observe the vibrations of the lattice of the solids, of which each mineral has its own characteristic frequencies. By tabulating the laboratory measured frequencies of different minerals and comparing the results with the astronomical infrared spectrum, we can identify the minerals. An alternative method of studying minerals is by X-ray techniques. When an X-ray strikes a mineral surface, it will kick some electrons off the atoms. Since different atoms require X-rays of different energies to eject an electron, one can identify the atoms in the solid by radiating the solid with a wide range of X-ray energies and observing the emergent spectrum after X-rays have passed through the solid. The way it is done is to employ synchrotron accelerators as X-ray sources and to let these high power X-rays pass through some known target material. The emergent X-ray after having passed the material will carry signatures of the material. For example, if a mineral is made up of iron, magnesium, and oxygen, these constituents will be revealed in the X-ray spectra.

Julia Lee, a young professor from Harvard University, has proposed to use this technique to study minerals in space. By pointing a satellite-based X-ray telescope to some known celestial X-ray sources, one can study the spectrum to see if there are any signatures of interstellar dust between the source and us. Lee was trained as an X-ray astronomer and has been using space-based X-ray telescopes to study black holes. However, in some of her X-ray spectra, she found features that are not arising from atoms and ions and therefore are unrelated to the blackholes. She concluded that such features are the result of interstellar iron solids between us and the blackhole. This was the beginning of the discipline of X-ray condensed matter astrophysics.

As we have learned in this chapter, gems and minerals are defined by the different kinds of atoms inside. If one can develop a catalog of X-ray spectra of known minerals, this information can be used to identify different kinds of minerals in space. In January 2012, Julia Lee organized an international workshop at the Radcliffe Institute of Advanced Studies.[1] Lee invited synchrotron physicists, geologists, space scientists, and astronomers from different countries to the workshop to formulate ideas on how to take this concept further. Obviously more laboratory work needs to be done with X-ray light sources, but X-ray telescopes with higher sensitivities and spectral resolutions are also required. Whether further breakthroughs in our understanding of stardust can be achieved through X-ray astronomy will rely on the construction of a new generation of space-based X-ray telescopes.

Not too long ago, many would have thought that the search for gems and minerals in space was nothing but fantasy. However, recent infrared spectroscopic studies of stars have shown that there are gems and minerals in space and we will detect a larger variety of them as we acquire better instruments and telescopes. This will not be easy as many gems contain rare elements and their low elemental abundances make them harder to detect. I firmly believe that the discipline of astromineralogy is here to stay and will flourish more in the coming decades.

A brief summary of this chapter
Stars can make minerals, including different forms of gems.

Key words and concepts in this chapter
- Gems
- Rubies, sapphires and spinels as minerals
- Gems in meteorites
- Identification of stellar gems by infrared spectroscopy
- X-ray spectroscopy of minerals

Questions to consider
1. When you visit a geology museum, you can see on display many different kinds of minerals, all having their own unique shapes and colors. How do you think these minerals are formed on Earth?

(continued)

---

[1] Radcliffe College used to be a women's college next to Harvard College in Cambridge, Massachusetts. After Radcliffe merged with Harvard in 1977, the campus of the College became the grounds for the Radcliffe Institute for Advanced Studies, the Harvard arm of a research think tank.

2. Some geologists find it difficult to believe that there are natural minerals in space, arguing that the astronomical evidence is "indirect". How reliable do you think the astronomical mineral identifications are? What are the possible sources of error?

Chapter 12
Diamonds in the Sky

People of all ages and cultures have always been fascinated by diamonds, attracted by their beauty, glitter, and rarity (and therefore value). For over 2000 years, diamonds were found mainly in pebbles in riverbeds. The discovery of diamond-containing rocks in South Africa led to the realization of volcanic pipes as a source of diamonds. It is interesting to note that diamond is a form of pure carbon, the fourth most common chemical element in the Universe, after hydrogen, helium, and oxygen. The rarity of diamonds is therefore not due to its chemical composition, but the result of how and where they are made. The diamonds on Earth are believed to have formed under high temperature (~1,000 °C) and pressure (4,500 times the pressure we experience everyday under the Earth's atmosphere) in the deep (100–300 km) interior of the Earth, and brought to the near surface by volcanic activities.

Chemically, diamond is a close cousin to graphite, a natural substance that is used to make pencils. Graphite is the stable form of carbon under low pressures, for example at the surface of the Earth. In fact, if you put a diamond in an oven, the precious diamond will decay into common graphite.

Besides its beauty, diamonds also have high utility value. It is the hardest natural substance known. In fact the name "diamond" is derived from the Greek word "adamas", which means "unconquerable". This physical property of diamond therefore leads to many industrial applications. For example, it can serve as a hard coating for machine tools and mining drills. The demand for diamonds for industrial use has created the need to synthesize diamonds artificially. In the short story "*The Diamond Maker*" published by H.G. Wells in 1911, he speculated on this possibility. The reality came in 1953 when small crystals of artificial diamonds were made under high pressure and high temperature, simulating the process of the creation of natural diamonds. A newer method is chemical vapor deposition (CVD), which creates diamond by vaporizing carbon and condensing the atoms on a cold substrate. Now diamonds are made routinely in the laboratory in a matter of days while nature needs millions of years. By 2006, 600 tons of synthetic diamonds were produced annually for industrial use, 20 times more than the amount of natural diamonds mined for gems.

S. Kwok, *Stardust*, Astronomers' Universe, DOI 10.1007/978-3-642-32802-2_12,
© Springer-Verlag Berlin Heidelberg 2013

Synthetic diamonds are moving beyond industrial use and are invading the realm of gems. After the collapse of the Soviet Union, the Institute of Geology in Novosibirsk in Siberia was short of money, so they lent their expertise in making diamonds to an American company to make synthetic diamonds for consumer use. The artificial diamonds are so good now that they are practically indistinguishable from the natural ones. Although they can still be distinguished in a laboratory by an expert, the difference is really only psychological because as far as visual appearance is concerned, they are identical to natural ones. It is impossible for the general consumer to tell the difference.

In addition to its ornamental and industrial values, diamonds have also served the role as indicators of past impact events (Chap. 3) which had significant effects on the evolution and development of living organisms on Earth. Diamonds of extraordinarily small sizes (nanometer scale) can also be synthesized through the process of explosion. It has been known since the 1960s that the ashes left behind from a TNT detonation contain tiny diamonds. They are given the name of nanodiamonds. Nanodiamonds have also been found in sedimentary rocks, and are usually associated with impact craters. Their existence has been hypothesized to be created from explosions as the result of external impacts. A recent study published in the *Proceedings of the National Academy of Sciences* identified nanodiamonds in a 13,000-year-old layer of sediment buried in the floor of Lake Cuitzeo in Mexico. The study suggests that a comet or asteroid entered the Earth's atmosphere at a shallow angle. The air shock caused by the resultant impact raised the surface temperature to a high value, resembling the conditions of TNT detonations, leading to the synthesis of nanodiamonds. This recent impact may have caused extensive environmental disruption, or even global cooling. The immediate aftermath may have resulted in the melting of surface rocks and wildfires with temperatures up to 900 °C. Long term effects could include the extinction of large North American animals such as the mammoths, mastodons, and dire wolves, or even causing a reduction in human population.

Because of the high abundance of carbon in the universe, the existence of interstellar diamond has long been speculated. It should be emphasized that diamonds are rare not because of their composition (they are made up of pure carbon after all), but the conditions under which they are made. If suitable conditions exist in space, diamonds could be present naturally in space. The first solid indication for extraterrestrial diamonds was the 1987 discovery of nanometer-sized diamonds in meteorites. This provided the first evidence that diamonds are not the sole domain of the Earth but in fact are present beyond the Earth.

Since meteorites are limited to the near-Earth environment, the search for the general presence of diamonds has to be done through remote spectroscopic observations. Since diamonds are nearly transparent in the visible and lack distinct features in the infrared, it is extremely difficult to detect interstellar diamonds by standard astronomical spectroscopic techniques. Furthermore, diamonds on Earth are formed under high temperature and pressure, conditions that are impossible to replicate in space. So, interstellar diamonds are conjectured to be formed in strong shocked gas, such as in supernova explosions. By the end of the twentieth century, the outlook for detecting diamond was bleak, both on observational and theoretical grounds.

The best way to search for diamond in space would be through an all-sky survey using the technique of infrared spectroscopy. By splitting the infrared colors very finely, astronomers can use the features present to determine the materials responsible for creating such features. This is done by comparing the astronomical spectra with the laboratory spectra of known substances and minerals. The first successful example was the detection of silicates mentioned in Chap. 6. Usually, astronomical infrared spectroscopic observations are targeted at specific stars or galaxies. This, however, would constrain our searches by our own biases. An all-sky search will suffer from no such problem. The *IRAS* satellite was the first to perform such a survey (Chap. 7). The LRS instrument would record a spectrum whenever the satellite scanned over a celestial object with sufficient brightness in the infrared. This is the best method to discover the unexpected.

When Kevin Volk and I analyzed the data from the *IRAS LRS* survey, we found that the majority of spectra could be assigned to minerals such as silicates and silicon carbide. Many also belong to the already famous UIR family of unidentified infrared features, now believed to be due to aromatic compounds (Chap. 10). However, there were a number of peculiar spectra that were completely different. Kevin Volk gave them the assignment of "U", standing for "unknown". In particular, several of them showed a previously unseen emission feature at the wavelength of 21 μm. This feature was so strong that it could not have been an instrumental effect. The only reason why such a strong emission feature was not found earlier is because *IRAS* was the first spectral survey in the mid-infrared wavelength region.

With the *United Kingdom Infrared Telescope*, we identified the optical counterparts of these sources, and I immediately recognized that these stars looked exactly the same as the proto-planetary nebulae that Bruce Hrivnak and I had been studying. In other words, this mysterious emission feature is produced during the proto-planetary nebulae phase of evolution (see Chap. 10). In Fig. 12.1 we show a picture of one of the 21-μm sources, the proto-planetary nebula *IRAS* 04296+3429, taken by us with the *Hubble Space Telescope*.

In December 1988, I presented these results at the European Space Agency meeting on infrared spectroscopy at Salamanca, Spain. We reported that four *IRAS* sources with no association with previously known astronomical objects have this unidentified emission feature. Using the *United Kingdom Infrared Telescope* and working with Tom Geballe of the Joint Astronomy Center in Hilo, Hawaii, we proceeded to look for new sources showing the 21-μm emission. Ten years later, when I reviewed this subject at the International Astronomical Union symposium on asymptotic giant branch stars in Montpellier, France, we had discovered 12 such objects.

Astronomers were totally puzzled by this discovery. Since the feature is strong, it has to be primarily made up of common elements such as hydrogen, oxygen, or carbon. By that time, we had already determined by ground-based optical spectroscopy that these stars are carbon-rich. Logically, we suspected that carbon is the major constituent of the carrier of this feature. But what form of carbon? Two forms of carbon have long been known to exist on Earth. One is graphite, the soft, black substance that we use for pencils; and the other is diamond, the hard, colorless stuff

Fig. 12.1 *Hubble Space Telescope* image of the proto-planetary nebula *IRAS* 04296+3429. The proto-planetary nebula *IRAS* 04296+3429 is one of the four sources first found to show the mysterious 21 μm emission feature in 1989. This *Hubble Space Telescope* image shows that it has a pair of bipolar lobes as well as a disk like structure in the equatorial regions. At this time, it is unclear which region in the nebula the 21 μm feature is being emitted

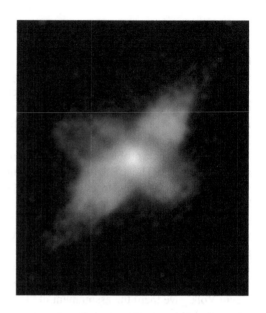

that we acquire at high cost to put around our fingers. However, the infrared spectra of neither graphite nor diamond show any resemblance to the 21-μm sources.

Nevertheless, very small diamonds have very different spectral characteristics from large, macroscopic diamonds. Louis d'Hendecourt of the University of Paris had been studying the infrared properties of nanodiamonds for some time. In 1996 during a conference in St. Louis, Missouri, H.G.M. Hill, a member of d'Hendecourt's group, saw our spectra of the 21-μm sources and noted a strong resemblance to his spectra of laboratory diamonds. In 1998, they published a letter in the journal *Astronomy and Astrophysics* suggesting that the 21-μm emission feature could be due to nanodiamonds.

Since all the indications are pointing to the carrier of the 21-μm feature being a carbon-based material, could it be another form of carbon other than graphite and diamonds? In 1995, Robert Curl, Harry Kroto, and Richard Smalley, in an effort to find the mechanism of forming long-chain molecules in the circumstellar envelopes of carbon stars, discovered a new form of carbon. They showed that 60 carbon atoms can be arranged in the form of a soccer ball-like cage and remain stable. This structure, with 32 faces consisting of 12 pentagons and 20 hexagons, was named buckminsterfullerene, or fullerene. In the popular press, C_{60} is known as buckyballs. (Chap. 9).

Could the 21 μm feature be due to fullerene? This possibility was investigated by Adrian Webster of the University of Edinburgh in Scotland. By performing theoretical calculations on the structures of fullerenes, he found that C_{60} molecules with hydrogen atoms attached to the corners could produce an emission feature similar to that observed by us. However, hydrogenated fullerenes (or fulleranes) have fine spectral features that are not seen in the astronomical spectra, so the 21 μm feature cannot be conclusively identified as fulleranes. Nevertheless, this remains a very attractive hypothesis.

Because of water absorption in the Earth's atmosphere, there are limits to the quality of infrared observations that can be carried out on a ground-based telescope. The launch of the *ISO* satellite in 1995 (see Chap. 11) gave us the opportunity to study the 21-μm feature in much greater detail. By comparing the *ISO* spectra of several 21-μm sources, we measured the peak wavelength of the feature to be at 20.1 μm. We were also able to determine the accurate shape of the emission feature and concluded that the carrier cannot be a small molecule, but a large molecular cluster or a solid.

Totally unaware of these astronomical developments, Gert von Helden of Matter's Institute for Plasma Physics in Nieuwegein in Holland was leading a team to synthesize titanium carbide (TiC) in the laboratory. They were able to make small clusters of TiC with 27–125 atoms. They then bombarded these small molecular units (called nanoclusters) with lasers. When the laser wavelength resonates with a TiC emission feature, the laser's photons cause the cluster to emit an electron. With this technique, they showed that TiC has a strong emission feature at 20.1 μm which coincides with the peak wavelength that we measured from our *ISO* observations. The TiC model was published in the journal *Science* in 2000 and received wide publicity.

The element titanium (named after the giant Titans in Greek mythology) is well known for its strength and light weight and is frequently used in aircraft and missiles. The molecule titanium oxide (TiO) is also commonly observed in the atmospheres of oxygen-rich asymptotic giant branch stars. So the suggestion of TiC as the carrier of the 21-μm feature is not unreasonable. There are, however, questions on whether there is a sufficient amount of titanium in these stars to explain the strong 21-um feature observed. After some strong arguments put forward by Aigen Li of the University of Arizona (now at the University of Missouri) and others on the question of abundance, the TiC nanocluster model gradually faded out of favor.

In 1996 at a conference titled "From Stardust to Planetesimals" held in Santa Clara, California, Thomas Geballe of the Joint Astronomy Centre in Hilo, Hawaii, presented a paper listing the four major mysteries in infrared spectroscopy. With the typical understatement of astronomers, he labeled this group of features as Classes A, B, C, and D. Class "A" is reserved for the UIR emission features seen in planetary nebulae, which we now know to be originating from aromatic compounds (Chap. 10). "Class B" features correspond to the spectra seen in proto-planetary nebulae, which include the 21-μm emission feature as well as features we now identify as due to aliphatic compounds. "Class C" designates a pair of infrared features at 3.43 and 3.54 μm seen in young stellar objects called Herbig Ae/Be stars. These two features were discovered by J.C. Blades of the *Anglo-Australian Observatory* and D.C.B. Whittet of University College London by accident, when they were searching for signatures of water ice in stars. This pair of features was observed by a variety of ground-based telescopes but their origin has remained unknown for over a decade. The best examples of stars showing these features are HD 97048 in the Chamaeleon dark cloud (Fig. 12.2) and Elias 1 in the Taurus dark cloud. Using *ISO*, a group of Belgian astronomers observed these two stars and

Fig. 12.2 Diamond is
detected in one of the stars
in the Chamaeleon Cloud.
The Chamaeleon dark cloud
is one of the most active star
formation sites near our solar
system. In the middle of this
picture is a reflection nebula.
Image credit: VLT, ESO

confirmed the ground-based results, but they were unable to offer any new insights into their origin. "Class D" is used by Geballe to refer to the UIR features in novae. Although Geballe did not try to assign chemical origins to these features, these qualitative descriptors do summarize the observational status very well. It allows astronomers to focus on the groups and attempt to find solutions to the mysteries.

Olivier Guillois of the Centre d'Etudes de Saclay in France had been studying the infrared properties of carbonaceous compounds, and took a strong interest in these astronomical results. In particular, he was curious about the pair of infrared features designated as Class C by Geballe. He did a search of the chemistry literature and found that a group of scientists at the Institute of Atomic and Molecular Sciences of the Academia Sinica in Taiwan has successfully taken infrared spectra of diamonds. The Taiwanese group led by Huan C. Chang irradiated synthetic diamonds with hydrogen. The hydrogen atoms then attached themselves to the corners of the diamond crystals. When they made infrared spectroscopic measurements of these diamonds, they noted that there are two distinct infrared features. Pure diamonds have very few infrared signatures, and these features are the result of the presence of hydrogen. Guillois noted that these two emission features are exactly at the wavelengths of the Class C astronomical features and concluded that they had found the answers to the mystery. Since there are lots of hydrogen atoms in interstellar clouds, it is not surprising that interstellar diamond would be covered with hydrogen. This identification was published in the *Astrophysical Journal Letters* by Guillois, Gilles Ledoux and Cécile Reynaud in 1999, and represents the first definite detection of diamond in space.

Guillois and his colleagues estimated that there are about 100,000 to 1,000,000 trillion (10^{17}–10^{18}) tons of diamonds in the surroundings of the star HD 97048. When we consider that on Earth we measure diamonds in units of carats (0.2 g), such a vast quantity of diamonds in a star is truly mind-boggling.

Is the chase for interstellar diamonds finally over? Based on the evidence that we have as of 2012, the case for the positive identification of diamonds (at least diamonds with hydrogen atoms attached) is very strong. However, there are lingering doubts. If diamonds are ubiquitous in the Galaxy, why is their presence not more widely detected among stars? It is true that the hydrogen-covered diamonds need the energy from starlight to excite them into emitting infrared light, so are the stars HD 9748 and Elias 1 unique? If the emission of these infrared features requires stringent radiation background, are the other diamonds just hidden from view because of their lack of ability to self-radiate? If this is the case, how can these hidden diamonds be discovered? In the next chapter, we will tell the story of how astronomers continued to search for these hidden diamonds.

A brief summary of this chapter
Stars can make diamonds in large quantities.

Key words and concepts in this chapter
- Diamonds and nanodiamonds
- Nanodiamonds as indicators of external impact and extinction events
- Diamonds in meteorites
- The 21-μm unidentified emission feature
- Carbon-rich proto-planetary nebulae as sources of the 21-μm feature
- Infrared signatures of hydrogenated diamonds
- Stellar synthesis of diamonds

Questions to think about
1. Diamonds are considered a rare commodity on Earth, but diamond is made up entirely of carbon, one of the most common elements in the Universe. What makes diamond rare on Earth?
2. Both diamonds and graphite are pure forms of carbon. Why is one so valuable and the other is not?
3. Thirteen thousand years ago can be considered as very recent in the context of the history of the Earth. Do you believe that an external impact in our recent past was responsible for the extinction of the large mammals in North America?

(continued)

4. External impacts have been suggested to cause major climate change. How significant are the climate changes caused by present human activities compared to celestial intervention of the past?
5. If diamonds are hard to make under terrestrial conditions, how do stars make diamonds in such large quantities?
6. Many of the discoveries in astronomy have been serendipitous discoveries. Why do you think that this is the case?
7. After over 20 years, the 21-μm feature has remained unidentified. Why do you think the identification is so difficult?

Chapter 13
A Mysterious Red Glow

Among living things, many organisms, ranging from bacteria to fish, exhibit bioluminescence. The most familiar example of bioluminescence to us is fireflies. Written record of this phenomenon goes back to 1000 BC. in the Chinese *Book of Odes* (*Shi Jing*). The Roman Army commander and philosopher Pliny the Elder (23–79 AD) recorded in his book *Naturalis Historia* descriptions of fireflies, glowworms, jellyfish, luminous mushrooms, and glowing wood. In Chinese folklore, there is a story about a diligent student in the Jin dynasty (265–420 AD) studying at night using the light of captured fireflies in a bag for illumination. Before the widespread use of artificial lighting in the twentieth century, natural light sources aroused great interest from ancient people. One can easily imagine the wonder and fear from the observations of a glowing rotten log, which was interpreted as coming from an evil spirit but we now know to have originated from luminous fungi. These natural light sources are different from other forms of light which are associated with heat, such as burning wood or candle flames. Aristotle referred these luminescence phenomena as "cold light".

With the beginning of ocean voyages in the fifteenth century came sailors' reports of the "burning seas". In 1492, Christopher Columbus reported the sighting of mysterious lights in the sea as his ships were approaching San Salvador. This phenomenon is common in the Caribbean waters and now traced to luminescent microorganisms. In 1634, English colonists saw light from fire beetles and thought the lights were from Spanish campfires, therefore deciding not to land in Cuba. However, bioluminescence is not limited to living organisms inhabiting the surface of oceans. Deep sea trawling reveals that many organisms living under the sea are self-luminous. Light probably offers an evolutionary advantage in the total darkness of the deep waters.

The first scientific study was made by the English scientist Robert Boyle (1627–1691) who in 1672 found that light emitted by rotting wood or flesh was due to glowworms. Robert Boyle is probably best known for the gas law that was named after him (Boyle's law), which states that a volume of gas decreases with increasing pressure if temperature is kept constant. He was a believer of alchemy, the practice of transforming metals into gold. In 1676, he reported to

S. Kwok, *Stardust*, Astronomers' Universe, DOI 10.1007/978-3-642-32802-2_13,
© Springer-Verlag Berlin Heidelberg 2013

the Royal Society his near success in changing silver into gold. In his study of bioluminescence, he was able to establish that light emission is inhibited by chemical agents such as alcohol and requires the presence of air. This demonstrated that bioluminescence is a chemical phenomenon.

We now know that jellyfish, squid, and phytoplankton all undergo bioluminescence. These natural fluorescent lights are not powered by electricity or external light, but by chemicals in their bodies. They use their ability to emit light either to lure prey or to scare away predators. But the reason why bacteria or fungi find the need to emit light is more difficult to understand. It is also not known why shrimp and squid shine on their own but not crabs or octopi.

It is therefore amazing that such a luminescence phenomenon can be happening in space. The discovery of stellar luminescence can be traced to the *Air Force Geophysical Laboratory (AFGL)* all sky rocket survey that we described in Chap. 5. Among the many interesting objects that were discovered in the survey, the object AFGL 915 surely must qualify as one of the most exotic. One of the major tasks after the rocket flights was to identify through ground-based observations the optical counterparts of the infrared sources found in the rocket observations. The U.S. Air Force was keen to catalog the natural objects in the sky that have infrared radiation. The rationale is that should intercontinental ballistic missiles (ICBMs) be launched by the Soviet Union, a sky monitoring system would be able to tell which infrared sources are incoming missiles and which are natural objects. Since the rocket-based telescopic observations could not pin point the locations of the detected infrared sources precisely, it was necessary to use a ground-based telescope to see which celestial source was actually associated with the AFGL infrared sources. These duties of ground-based observations were taken up by two astronomers: Martin Cohen from the University of Minnesota and Mike Merrill of the University of California at San Diego. Martin Cohen was a British student who went to the U.S. to learn the techniques of infrared astronomy and Merrill was a student at UCSD. Since the University of Minnesota and UCSD jointly operated the *Mt. Lemmon Telescope*, it was natural for Cohen and Merrill to work together. With an all-sky survey in two visible colors (red and blue) already in existence at that time in the form of the *National Geographic Society-Palomar Observatory Sky Survey*, their first job was to see whether there was any optical counterpart for the infrared source on the Palomar plates. On the position of object number 915 in the *AFGL* survey, they saw on the Palomar red plate a most unusual neatly rectangular object; they named it the "Red Rectangle". In November, 1973, they took a better picture of the object using the *Kitt Peak National Observatory* 4-m telescope in Arizona and found it to have a set of spikes in the form of an "X". In a more recent picture taken by the *Hubble Space Telescope*, these four spikes seem to form the basic structure of a spider web (Fig. 13.1).

The Red Rectangle is not only most unusual in its shape, but also in its spectrum. In 1979, Martin Cohen was observing at the 3 m *Shane Telescope* at *Lick Observatory* in California. When he took an optical spectrum of the object, he was surprised to find an "enormous bump" in red color. This excess red color originates from the nebula, totally unlike the star that illuminates the nebula. This discovery was so

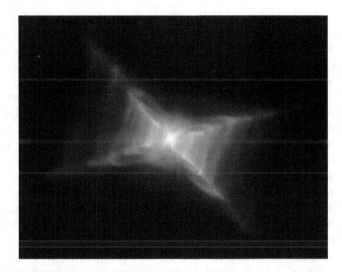

Fig. 13.1 *Hubble Space Telescope* image of the Red Rectangle.
Hubble Space Telescope image of the Red Rectangle. Spider web-like structures can be seen spanning the X-shape structure. The star responsible for exciting the ERE can be seen in the middle. Image credit: NASA/ESA, Hans Van Winckel (Catholic University of Leuven, Belgium) and Martin Cohen (University of California, USA)

astonishing and unexpected that he only made a very cautious interpretation in his publication of the results in the *Astrophysical Journal* in 1980.

This red emission could be an isolated anomaly in a peculiar object but this is not the case. Systematic observations of galactic reflection nebulae by Adolf Witt of the University of Toledo in Ohio revealed that this phenomenon (which he calls "extended red emission", or ERE for short) is wide-spread in the Galaxy. Not only is it observed in reflection nebulae, it is also seen in planetary nebulae, cirrus clouds high above the galactic plane, and even in external galaxies. Adolf Witt was able to make a connection between these diverse objects: they all have dust particles and an excess of ultraviolet light. This led him to suspect that ERE is related to the physical process of photoluminescence.

We all know that some solid substances (called phosphors) glow in the dark after having being placed under the Sun for some duration. Our understanding of this phenomenon began with the work of the German physicist Johann Wilhelm Ritter (1776–1810). Ritter was born in Silesia, which used to be Prussian territory but is now part of Poland. After hearing about Herschel's detection of heat ray (now known as infrared light) in 1800 (Chap. 5), Ritter experimented with silver chloride, a compound that turns black when exposed to sunlight. Chemical reactions that occur as the result of exposure to light are the basic principles behind photography. In order to test how these reactions depend on the color of light, he passed sunlight through a glass prism and placed silver chloride under different colors broken up by the prism. He found that the reactions are stronger (i.e., silver chloride darkens more) in blue light than in red light. When he put the compound beyond the extreme violet end where no sunlight could be visibly observed, the silver chloride reaction became even stronger. Ritter proved that there is energy beyond the violet light. This energy was labeled "chemical rays" as it causes chemical reactions. We now realize that there is no intrinsic difference between this "chemical ray" and the other visible colors, and we now call this "ultraviolet light".

In 1801, Ritter experimented with the behavior of phosphors under light of different colors. To his surprise, he found that phosphors luminesce brightly after being exposed to light beyond the color violet. He therefore discovered that some materials can convert ultraviolet light into visible light. Nowadays, we apply this principle in the use of fluorescent lamps where a phosphor coating is put inside a gas tube filled with argon and mercury vapor. When electricity is applied, the gas emits ultraviolet radiation which is absorbed by the phosphor coating and the energy is converted into visible light. Unlike professional scientists of today, Ritter had no regular income to support his research. He died young (age 33) and poor and did not live to see that his discovery led to the practical invention of the fluorescent lamp.

When we study the properties of the Red Rectangle, we find that its self-radiation in the red color is based on the same principle. Through an unknown material it contains, the Red Rectangle can efficiently convert the ultraviolet light emitted from its central star into red-colored light that we see in the nebula. We can therefore consider the Red Rectangle our fluorescent lamp in the sky.

Although inorganic phosphors such as zinc sulfide are commonly used as industrial luminescent materials, it is interesting to note that many organic compounds also have this capability and are widely used in fluorescent paints and dyes. Organic phosphors are particularly preferred in the use as invisible markers, as they often absorb no visible light (therefore "invisible") and only show fluorescence under ultraviolet light. Next time you play an investigator and suspect a hidden message somewhere, make sure that you carry an ultraviolet flashlight.

So what is the substance inside the Red Rectangle that gives it its photoluminescence? It has been 30 years since its discovery and after many debates among scientists there is still no agreed answer. Since ERE is seen throughout the Galaxy, it must be based on an element which is common. The correlation of its existence with the general carbon richness of the nebulae suggests that the substance is carbonaceous. But this is as far as astronomers can agree. There is no known substance that can match the efficiency and specific colors of the observed ERE. At the end of the twentieth century, two separate groups in the U.S. and France independently came up with the idea that the ERE could arise from silicon nanoparticles. Everyone got excited for a while, but in the end it did not work out.

Diamonds have long been known to undergo fluorescence under ultraviolet light. Could diamonds play a role in the creation of the ERE? In Chap. 12 we mentioned that very small diamonds are probably prevalent in space, as they are seen in large quantities in meteorites. In 2006, a Taiwanese team led by Huan-Cheng Chang bombarded nanometer-size diamonds with a high-energy proton beam from a particle accelerator, and then subjected the nanodiamonds to high heat at 800° C in order to simulate space conditions. These bombardments created defects in the diamonds, and when they shone yellow and blue light on the diamonds, red light emerged.

Nanodiamonds are so small that you have to lay 1,000 of them side by side to equal the width of a human hair or pile up 30 trillion of them to equal the weight of a half-carat diamond in a typical engagement ring. Unlike terrestrial diamonds, which are made at great depths of the Earth's crust under extremely high temperature and pressure and brought to the surface by volcanic activities, artificial nanodiamonds can be synthesized under low pressure using chemical vapor deposition techniques. These conditions are not dissimilar to the physical conditions in the stellar winds of old stars. Chang and I suggested that these diamonds are made in the surroundings of old stars like the Red Rectangle and NGC 7027 (Fig. 10.1) and are subsequently ejected in the interstellar space by their stellar wind. These diamonds then travel through the Galaxy, undergo luminescence by absorbing diffuse starlight in the Galaxy. The presolar nebula may have also inherited some of these stellar diamonds. The fact that nanodiamonds are recovered in meteorites lends support to this hypothesis.

We first presented these results at the American Astronomical Society meeting in Washington in January 2006 showing an excellent match of the fluorescent spectrum of nanodiamonds with the astronomical spectrum of ERE. The full results were published 2 months later in the *Astrophysical Journal*. If ERE indeed originates from diamonds, then diamonds must be everywhere in the Galaxy as ERE is seen throughout the Galaxy. Rather than a rare commodity as it is regarded on Earth, diamonds may be the most common thing in the Galaxy. If this is true, then diamonds are fluorescent lamps in the Galaxy.

The most mysterious aspect of the ERE phenomenon is that it is everywhere. For a substance to shine throughout interstellar space and be seen in distant galaxies, this must be a very common and not an exotic substance. Since we believe that nuclear reactions inside stars are the ultimate source of energy, there must be an efficient way of converting stellar energy output to this red glow in interstellar space. Although the ERE does not have the household name recognition of "dark matter" or "dark energy", it is certainly more real and no less important.

As of 2012, there is no consensus on the origin of the ERE. In fact, a special session was organized at the General Assembly of the International Astronomical Union in August of 2012 in Beijing to discuss this topic. There is no doubt, however, that the material that emits the ERE is a carbon-based material as only carbon has the abundance and chemical diversity to create this photoluminescence that is seen all over the Galaxy. It is also possible that the ERE is caused by an organic compound. This is not an unreasonable supposition as the UIR bands, almost certainly due to a complex organic compound, are also widely seen through-out the Galaxy. Are these two phenomena tied in any way?

A brief summary of this chapter

A mysterious optical phenomenon called the extended red emission remains unexplained. It could be due to diamonds or other forms of organic materials.

Key words and concepts in this chapter

- Extended red emission
- The Red Rectangle
- Photoluminescence and bioluminescence
- The fluorescent properties of nanodiamonds
- Fluorescent lamps of the Galaxy

Questions to think about

1. The mysterious green line in planetary nebulae, first attributed to an unknown element "nebulium", was later found to be due to the ordinary atom oxygen (Chap. 8). This was achieved as a result of the development of quantum physics, and the realization that the different conditions in space make atoms behave differently. There are still many unsolved spectroscopic astronomical phenomena whose origins have not been identified. Examples include the UIR features (Chap. 10), the 220 nm ultraviolet feature (Chap. 10), the ERE, and the 21 μm emission feature (Chap. 12). Why are these spectral features so difficult to identify? Could they originate from similar substances?

Chapter 14
A Celestial Origin for Oil?

Since the industrial revolution, we have been increasingly relying on fossil fuels—coal, oil, and natural gas—as our primary source of energy, replacing the burning of wood and the use of animal power. Gasoline, diesel and jet fuel (all refined products of petroleum) are used to propel cars, trucks and airplanes, natural gas and oil to heat our homes, coal to generate electricity. The role of these energy sources is so important that wars have been fought to gain control of them, and the drive to gain access to energy is believed to be a major force of modern international economics and politics.

Different forms of oil have indeed been used as energy source for at least several hundred years, or may be even thousands of years. Ancient people may have used easily accessible oil exposed in river banks for burning. Shale oil, which is a form of kerogen extracted from rocks, was recorded to have been used to light lamps as early as the seventeenth century. Shale oil production in fact became widespread in the world in the eighteenth and nineteenth centuries. By the mid-nineteenth century, crude oil was discovered in the Middle East. It is usually found in reservoirs in sedimentary rocks. Sedimentary rocks are formed by sand and clay from land surfaces eroded by moving water and deposited in layers. The oldest rock containing oil can be traced back to the Cambrian Period about 500 million years ago. Crude oil is a black or brownish viscous liquid consisting of a mixture of different hydrocarbons. Oil usually flows easily but heavier forms of oil can be in a semi-solid form. Since crude oil is easy to extract and burns better, it has become the dominant energy source by the twentieth century.

At the beginning, the use of petroleum was limited for use of lighting. Kerosene, the distilled product of petroleum, was used as fuel for lighting lamp. Kerosene gradually lost its popularity when electric light came into use in the early twentieth century. The use of petroleum then shifted to the transportation sector. With the rising use of internal combustion engines and their deployment in motor vehicles, the demand for petroleum sky rocketed. Petroleum derivatives are also used for paving roads, and they are the basis of many plastic products. They are also found in personal care products such as petroleum jelly and in various medicines. Our modern life style is inseparable from the use of petroleum.

S. Kwok, *Stardust*, Astronomers' Universe, DOI 10.1007/978-3-642-32802-2_14,
© Springer-Verlag Berlin Heidelberg 2013

In terms of chemical structures, coal (a solid), petroleum (a liquid) and natural gas (a gas) all belong to the family of hydrocarbons and they are made up primarily of hydrogen and carbon atoms. Natural gas consists of mainly simple gaseous molecules such as methane, but petroleum is a complex mixture of chains of hydrocarbons (see Chap. 10). Coal (Fig. 10.6) has an even more complex and irregular structure, being a mixture of rings and chains. Sometimes oil is found in tar sands and oil shales. Methane can also be in a crystal solid form enclosed in water or ice (called methane hydrate). For example, scientists have been looking for methane hydrate under the Canadian Arctic.

In spite of the enormous practical importance, the origin of these fuels is not completely understood from a scientific point of view. Fossil fuels, as the name implies, are believed to be products broken down from plant and animal debris deposited near the surface of the Earth. They are generally found in sedimentary basins and along continental shelves. The theory of the biological origin of hydrocarbons goes back more than one hundred years to the 1870s. In 1934, Alfred Treibs (1899–1983) found porphyrins, a class of complex aromatic rings containing nitrogen, in petroleum. The discovery of nickel- and vanadium-porphyrin complexes in petroleum suggests a link to iron–porphyrin in hemoglobin in the blood of animals and magnesium-containing chlorophyll in the leaves of plants, therefore giving the biological theory its first experimental support.

At that time, the Earth was thought to have condensed from a hot molten sphere of rock, and therefore no primordial hydrocarbons could have survived. Instead, crude oil and natural gas that we find today were believed to have decayed from remains of vegetation and animals over several hundred million years.

The major biological sources are believed to be single-celled planktonic plants such as blue-green algae, and single-celled planktonic animals such as foraminifera. These organisms were common in the ancient seas, rivers, and lakes and their remains were mixed with sediments buried in the bottom of seas. The heat and pressure of the Earth first cooked these remains into a waxy substance called kerogen, which in turn was broken down into oil and natural gas. The insoluble kerogen remains in the rocks whereas the oil released migrates through pores and capillaries upward until it is trapped by a barrier. The accumulated oil forms a reservoir, which is generally referred to as an oil field. Since plants and animals only exist near the surface of the Earth, oil and gas deposits were consequently expected to be confined to the Earth's crust.

How much oil is there on Earth? In 2000, the U.S. Geological Survey estimated that there are about three trillion barrels of oil which are recoverable, of which about 710 billion barrels have already been consumed by 1995. The rate of world oil consumption in 2007 was 30 billion barrels per year. The consumption rate is expected to rise as countries like China and India develop. Although the amount of reserves is uncertain, it is quite likely that oil supplies will be exhausted in decades.

However, not all hydrocarbons on Earth are of biological origin. The discovery of methane in hydrothermal vents shows that at least some hydrocarbons are abiogenic—that is, not originating from plant or animal remains. These hydrocarbons are believed to be formed from carbon dioxide through water–rock interactions. The more important question is: are these abiological hydrocarbons present in large quantities in the Earth's crust? In order to determine their global quantity, one has to be able to separate them from biological hydrocarbons. By studying the gas discharges from boreholes in hard rock mines operating in the Canadian Shield, Barbara Sherwood Lollar of the University of Toronto in Canada concluded that they do not represent a significant fraction of the total hydrocarbon reserves.

Although the idea that coal, oil, and natural gas descended from former living things is firmly entrenched in the Western geological community, it is not accepted by all. For example, the biological origin of oil was questioned by Hoyle in his book *Frontier of Astronomy*:

> "The idea that oil, so important to our modern civilization, has been squeezed out of the Earth's interior derives an immediate plausibility from Urey's discovery that the meteorites contain small concentrations of hydrocarbons. The presence of hydrocarbons in the bodies out of which the Earth is formed would certainly make the Earth's interior contain vastly more oil than could ever be produced from decayed fish – a strange theory that has been vogue for many years" (Hoyle 1955, p. 37).

Hoyle's challenge to the tradition was followed up by Thomas Gold (1920–2004) (Fig. 14.1), a professor at Cornell University in New York, who suggested in 1977 that the hydrocarbons are extraterrestrial in origin and petroleum, in particular natural gas, can be found in great depths. His theory is based on the recent realization that the Earth was never all molten but was built up from solids. Through the accretion of dust and rocks in the primordial solar nebula, Earth was created as a minor entity of the Solar System four and half billion years ago. If hydrocarbon molecules were already present in the solar nebula, they would be part of the Earth. Because of their lighter weight, these hydrocarbons would gradually float upwards to the surface. They would be seen as plumes of flammable gas, or if trapped inside porous rocks, as wells of oil and gas retrievable by drilling.

Gold's theory, if true, would have major practical implications. The oil and gas reserves would be much, much larger than the current estimates. It would also affect exploration strategies, because oil and gas would not be confined to sedimentary rocks in the uppermost layers of the Earth's crust, but would also be in reservoirs hundreds of kilometers deep. On the question of supply, a deep well under high pressure can contain much more gas than a shallow one, and an oil well, if maintained by continuous flow from below, would be practically inexhaustible. To test this theory, Gold was able to convince the Swedish government to drill two deep wells, Gravberg 1 in 1986–1990 and Stenberg 1 in 1991–1992. However, no conclusive evidence for deep oil was found.

Although less well known in the West, the abiogenic theory of the origin of oil has been discussed widely in the former Soviet Union since the 1950s. Based on geological arguments, the Russian geologist Nikolai Alexandrovitch Kudryavtsev proposed in 1951 that the Athabasca Tar Sands in Alberta, Canada, originated from a high, deep volume which cannot be explained by biological remains. Tar sands are a mixture of sand, clay, water, and a viscous form of petroleum. It is different

Fig. 14.1 Tommy Gold.
Thomas Gold was born in Vienna, Austria. As a student in Cambridge at the beginning the Second World War, he was interned as an enemy alien first in England and then in Canada. After his release from internment and return to Cambridge, he did work on radar where he met Fred Hoyle and Hermann Bondi. After the war, the three scientists developed the Steady State theory on the origin of the universe. He made many contributions to astronomy, most notably his theory of pulsars. He also generated many controversies; his theory that the Moon is covered with a layer of dust infuriated many lunar scientists. His popularization of the abiogenic theory of oil was also highly resented by geologists

from crude oil in the sense that tar sands are in a semi-solid state and are so thick that they cannot be extracted through the conventional way as in common oil wells. In order to extract the petroleum content, tar sands have to be heated by steam, which greatly increases the cost. Instead of remains of living organisms, Kudryavtsev suggested that these hydrocarbons are geological in origin and they just migrate to the surface of the Earth due to their lower density. Arguments have been put forward by both sides for and against the biological origin of oil, with the geologists (at least those in the West) solidly behind the biogenic theory.

If the hydrocarbons are not biological in origin, where did they come from? Gold envisioned that these molecules were made in interstellar clouds and carried to Earth by meteoroids or comets. Although there was no empirical evidence for this hypothesis at the time, recent astronomical observations have given a new dose of credence to Gold's theory. Old stars near the end of their lives are found to be able to manufacture hydrocarbons with a degree of complexity not imagined before. These stars, acting like chimneys, pollute the galactic environment by releasing large quantities of these organic compounds (Chap. 10). Since the Solar System was created from an interstellar cloud, the Earth might have inherited some of these materials.

Fossil fuels are primarily made up of two elements: carbon and hydrogen. While hydrogen was produced in the Big Bang during the early days of the universe, all carbon atoms, including every one of those in our bodies, were made in stars (Chap. 4). As a star like our Sun gets old, it becomes brighter, larger, and redder. Its central core also becomes denser and hotter, eventually hot enough to ignite the helium atoms and burn them into carbon through nuclear reactions. In the last one million years of a star's life, these carbon atoms are brought to the surface making

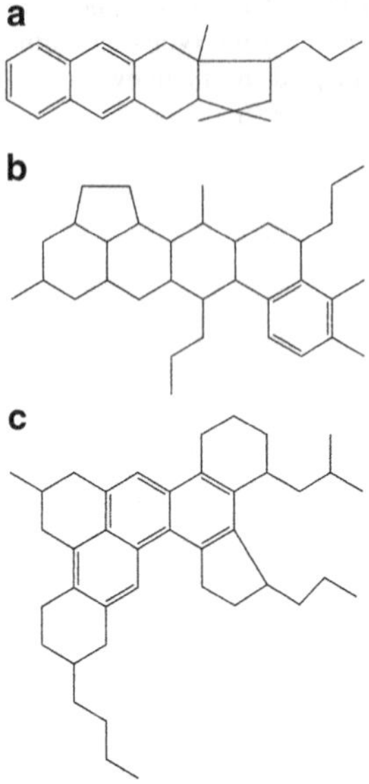

Fig. 14.2 Examples of
petroleum fractions.
Molecular structures of some
examples of petroleum
fraction that can account for
the infrared features observed
in proto-planetary nebulae.
The corners of each line
represent a carbon atom. The
hydrogen atoms attached to
the carbon atoms are not
shown. These petroleum
fractions are a mixture of
aromatic (ring) and aliphatic
(chain) structures.
Figure adapted from Cataldo
et al. (2002), *International
Journal of Astrobiology*, 1,
79–86

the stellar atmosphere rich in carbon. In Chap. 10, we learned that old stars are capable of making complex carbonaceous organic compounds, with structures similar to those of natural primitive coal, or artificial substances such as HAC or QCC. In 2002, Franco Cataldo of the Soc Lupi Chemical Research Institute in Rome, Italy, extracted some molecular fractions from petroleum (Fig. 14.2) and compared the infrared spectra of these extracts with the astronomical spectra of proto-planetary nebulae. Surprisingly, he found a close match. In several papers published in the *International Journal of Astrobiology*, he championed the idea that the unidentified infrared (UIR) bands observed in the circumstellar and interstellar space are due to petroleum. He does not believe that these petroleum fractions are of biological origin. It is more plausible that old stars are able to synthesize oil naturally without involving living organisms.

Whatever the exact makeup of these organic compounds is, there is no doubt that dying stars are making organic matter with a degree of complexity not imagined by astronomers just a few years ago. The creation of complex organic molecules is no longer the sole domain of the Earth.

Before one decides on the origin of hydrocarbons on Earth, we should take a look at the inventory of carbon on Earth. Carbon can be in many different forms and a breakdown of their presence on Earth is given in Table 14.1. This table breaks down the distribution of carbon among the different components of the Earth: the solid crust of the Earth (the lithosphere), the oceans (the hydrosphere), the atmosphere, and the biosphere. There is carbon dioxide in the atmosphere.[1] When it is

---

[1] There is a lot of interest in carbon dioxide in the atmosphere these days because it is a greenhouse gas responsible for the warming of the Earth.

Table 14.1 Carbon pools in the major reservoirs on Earth (adapted from Falkowski et al. 2000)

Pools	Fractional quantity (Giga tons)
Atmosphere	720
Oceans	38,400
Total inorganic	37,400
Surface layer	670
Deep layer	36,730
Total organic	1,000
Lithosphere	
Sedimentary carbonates	>60,000,000
Kerogens	15,000,000
Terrestrial biosphere (total)	2,000
Living biomass	600–1,000
Dead biomass	1,200
Aquatic biosphere	1–2
Fossil fuels	4,130
Coal	3,510
Oil	230
Gas	140
Other (peat)	250

dissolved in rainwater, it is carried to the sea; so oceans are also a reservoir of carbon. Chemical reactions with calcium produce calcium carbonate, which is stable and accumulates in sedimentary rocks. Carbonates are therefore the largest reservoir of carbon on Earth, estimated to contain over 60,000 trillion tons. Next is carbon in the organic form of kerogen, estimated to be about 15,000 trillion tons. In comparison, all biomass on Earth, including plants and animals, account for less than one trillion tons of carbon. The fossil fuels coal, oil, and gas add up to another four trillion tons. So if kerogen is the remnant of dead living things, it would have taken a long while to accumulate.

Thomas Gold's theory of the abiogenic origin of oil is highly controversial and is not accepted by the geological community. One of the objections is that primordial hydrocarbons are unlikely to have survived the high temperatures and shocks generated in the process of accretion during the formation of the Earth. During the period of planet formation, the zone occupied by the terrestrial planets is expected to have been hot, and the interstellar grains in this zone are likely to have lost not only their icy mantles but also their organic contents.

High temperature is not the only problem. Collisions may also have affected the Earth's ability to retain its primordial organic substance. According to current thinking, Earth was formed in a hot, largely molten state after a sequence of coalescence of relatively large bodies (embryonic planets, not just 10-km size planetesimals). The last of such encounters between the Earth and a large body (estimated to be about the size of Mars) resulted in the creation of the Moon four-and-half billion years ago.

Although we take the Moon for granted as one of the two major sources of light in the sky, its existence posed a great problem for scientists. Since the dawn of scientific thinking, the Moon has been variously thought to have been captured by the Earth, formed out of the same Sun-orbiting material that created the Earth, and broken off the Earth as the result of rotation. All these theories have great difficulties, as the Moon has a very different composition from the Earth's. While the Earth has an iron core that constitutes one third of its mass, the Moon has a much lower average density, and no more than 4 % of its mass is in the form of iron. The most popular hypothesis today is the "giant impact theory" which suggests that the Moon was formed out of the debris of a collision between the Earth and a Mars-sized proto-planet. These debris were then trapped by the gravity of the Earth, circulated around the Earth, collided with each other, and gradually aggregated into a solid body in the form of the present Moon. Since such impact mostly stripped off the outer layers of the Earth, the Moon did not inherit the iron content which is deep inside the Earth.

This collision probably deprived the Earth of its primordial atmosphere and evaporated all its surface water. It is likely that most of the surface organic compounds were also gone. In Gold's original "deep-earth gas theory", he only considered simple hydrocarbons as the primordial gas. At that time, he did not have the knowledge that organic matter as complex as kerogen could be produced by stars. Solids like kerogen are certainly more resilient and had a greater chance of survival during the early days of Earth formation.

It is now known that carbonaceous chondrites, the most primitive objects in the Solar System, contain a significant amount of insoluble organic matter (Chap. 15). Laboratory analysis of the insoluble organic matter extracted from the Orgueil, Murchison, and Tagish Lake meteorites has revealed that this organic matter consists of a network of aromatic rings linked by short aliphatic chains. Such structures are extremely similar to the carbonaceous organic compounds synthesized by proto-planetary nebulae (Chap. 10). Since the early Earth suffered from large-scale mass impacts, micrometeorites probably delivered a considerable amount of kerogen-rich material to the terrestrial oceanic crust. Having accumulated on the ocean floor and been trapped in sediments, these materials would develop into oil reserves just as dead plankton would according to the biogenic origin of oil.

In order to explain the existence of coal, Gold hypothesized that micro-organisms deep inside the Earth may play a role in converting hydrocarbons into coal through hydrogen extraction. If kerogen is the primordial material instead, then this hypothesis is unnecessary. Coal is known to have evolved from kerogen through the natural loss of hydrogen and nitrogen. Petroleum is often found in sedimentary rocks rich in kerogen and is believed to be naturally degraded from kerogen over time. Both processes are accompanied by a release of natural gas.

Is there enough celestial organic matter to account for the known reserves of coal, oil, and gas? From Chap. 10, we learned that complex hydrocarbons are routinely produced by stars and ejected in large quantities into interstellar space. Our best estimate is that carbon stars eject the equivalent of the mass of the Sun (300,000 times the mass of the Earth) into the Galaxy every year. Even if a small fraction of that is in the form of organic matter such as kerogen or HAC, over the several billion years' lifetime of the Galaxy, the total accumulated amount is considerable (over a trillion Earth mass). The solar nebula, from which our Solar System was formed, was likely to have contained remnants (may be of the order of 1,000 Earth mass) of these organic compounds. Whether these compounds could have survived the creation of Earth and remained intact deep inside the Earth, is uncertain. We must remember that although the Earth seems immense to us, it is but a speck of dust on the Galactic scale. In terms of chemical composition, the Earth is similar to the kind of silicate dust that is routinely produced by old stars rich in oxygen. Since the total amount of carbon contained in kerogen, coal, oil and gas in the Earth's crust is about one millionth (10^{-6}) of the Earth mass, even a small amount of extraterrestrial organic matter could easily overwhelm our known fossil fuel deposits.

If oil on Earth could have descended from extraterrestrial materials, why couldn't the same process lead to oil reservoirs in other planets? The *Cassini mission* found that Saturn's moon Titan contains a huge amount of hydrocarbons (Chap. 15). Incredible as it may seem, there could also be oil and gas on Mars. Like Earth, Mars underwent tectonic processes that could lead to the trapping of hydrocarbons in sediments. In 2006, Susana Direito and Maria Webb of Portugal proposed that we could seriously explore the presence of oil on Mars. Using data from the Mars Orbital Camera of NASA's *Mars Global Surveyor* spacecraft, they searched for the evidence of oil seeps from dark features in the images. For example, they suggested that the dark features found on the Martian South Pole could be due to oil on ice.

As remote as these possibilities are, we cannot help but ponder their implications. Whether natural gas was made at the Earth's crust and later sank into the Earth, or rose from deep inside the Earth greatly affects the exploration strategy. The belief that there is a finite supply of oil plays a significant role in the economic planning of nations. Diplomatic alliances are made and wars fought in order to secure energy supplies. Given the important consequences, how can we afford not to look further into this question? At the very least, we ought to pay more attention to the relations between the two disciplines that study the heavens and earth. Astronomers and geologists have much to learn from each other.

A brief summary of this chapter
A discussion on the theories (biological and non-biological) theories of the origin of fossil fuels.

Key words and concepts in this chapter
- Coal, oil, and natural gas as fossil fuels
- Oil as decay products of plant and animal debris
- Abiogenic theory of the origin of oil
- Carbon reserves on Earth
- Meteorites and insoluble organic matter
- Hydrocarbons in planetary satellites

- Stars as sources of kerogen-like materials
- Possible contribution of hydrocarbons from extraterrestrial sources

Questions to think about
1. For thousands of years humans relied on the chemical energy released from burning wood as a source of heat, and the biological energy extracted from animals (horses, oxen, donkeys, etc.) for kinetic power. In the last 200 years, the chemical energy stored in coal and oil have become the major sources of energy. Nuclear energy based on the splitting of atoms (fission) has been used since the 1950s. What do you think could be the energy sources for the future?
2. It has been suggested our rising consumption of oil is unsustainable. There are also other suggestions that the demand for oil can be met by continued discoveries of new reserves. Which do you think is correct?
3. Thomas Gold hypothesizes a large reservoir of oil deep inside the Earth. Why have we not explored the interior of the Earth for sources of oil?

Chapter 15
Organics in Our Solar System

Our Earth is full of organic molecules and compounds as a result of life. Are there similar organics elsewhere in the Solar System? There are two ways that we can find out. One is to perform remote observations—for example through spectroscopic observations using telescopes—and the other is by actually collecting samples and analyzing them in the laboratory. Astronomical observations usually rely on spectroscopic observations to detect the spectral signatures of the organic molecules. This technique has been applied to detect molecules (Chap. 9) as well as organic solids (Chap. 10) in stars and in the interstellar medium. However, the proximity of Solar System objects offers another avenue of investigation, and that is by experimentation. Instead of just passive observations, the development of space travel allows the possibility of actually visiting Solar System objects to perform in-situ observations, and even collecting samples and bringing them back to Earth for closer inspection. The possibility of sample manipulations allows the use of much more powerful techniques to study the composition of the samples and more definite conclusions about their chemical structures.

Although some of the spectral signatures of organic molecules were detected in the planets Jupiter and Saturn as early as 1905, they were not recognized as being due to methane and ammonia until 1932. In 1944, some optical spectral signatures of methane were discovered in the spectra of Uranus, Neptune, and Titan, establishing that some organic molecules are present in planets and their satellites. With the invention of infrared techniques, our ability to search for organic molecules in the Solar System greatly expanded in the 1970s. Steven Ridgway of *Kitt Peak National Observatory* was among the first to employ this new technique and was able to discover a large host of organic molecules in Jupiter. The *Galileo mission* was launched on October 18, 1989 by the Space Shuttle Atlantis and arrived at Jupiter on December 7, 1995 after having received two episodes of boost in speed from the gravitational fields of Venus and Earth through flying near these planets. Before the spacecraft reached Jupiter, it released an entry probe into the atmosphere of Jupiter in July 1995. The *Galileo probe* descended through 150 km of the Jovian atmosphere and its six instruments collected 58 min of data during the descent. Through the entry probe and

S. Kwok, *Stardust*, Astronomers' Universe, DOI 10.1007/978-3-642-32802-2_15,
© Springer-Verlag Berlin Heidelberg 2013

remote observations from the orbiter, the *Galileo mission* extensively studied the planet Jupiter and found organic molecules to be common in the clouds and haze in the Jovian atmosphere. Through subsequent astronomical ground-based and satellite-based observations, organic molecules have been found in the atmospheres of the planets Jupiter, Saturn, Uranus, Neptune, and Pluto, as well as satellites such as Titan and Triton.

Sample retrieval is much more difficult, but we have managed to collect samples from the Moon, Mars, comets, and asteroids by sending spacecrafts to these objects and analyzing their contents with on-board instruments. However, the most efficient way to explore the chemical content of the Solar System is to make use of materials that naturally fall onto the Earth, namely meteorites and interplanetary dust particles.

The common perception of meteorites is that they are rocks. It is true that the majority of meteorites are made of metals or minerals. These meteorites are called iron meteorites and stony meteorites, respectively, and they are the result of heating, melting, mixing, and condensation after the formation of the Solar System. However, there is a rarer class of meteorites called carbonaceous chondrites that are believed to be pristine and untouched by the cooking process during the early days of the Solar System. As we can tell from their name, carbonaceous chondrites are rich in carbon.

The first evidence for the presence of organics in meteorites was found in the Orgueil meteorite which fell in 1864 (Chap. 2). Chemical analysis of the meteorite done soon after the fall showed that this meteorite was different from others in that it contained a substantial amount of carbonaceous materials. Interest in this peculiar meteorite was renewed in the mid-twentieth century when analysis techniques became much more sophisticated. The development of gas chromatography and mass spectrometry allowed specific organic components to be identified. Chromatography, which means "color writing", is a technique used by chemists to separate a mixture of volatile components, therefore allowing the identification of each of the components. Mass spectrometry, on the other hand, relies on the different mass-to-charge ratios of the components and separates them based on their different paths of motion under the influence of electric and magnetic fields. The miniaturization of computers has allowed the integration of these two techniques into stand-alone gas chromatography-mass spectrometry (GC-MS) machines, which are now commonly used for forensic identification and are in fact employed in airport security for the detection of specific substances carried by passengers. It is the development of these forensic techniques that has allowed scientific detectives to realize the degree of richness in organics in meteorites.

Although these discoveries made a great impression on scientists working on meteorites, there was, however, initially great reluctance in the general scientific community to accept these results. The presence of complex organics in an extra-terrestrial object seemed too fantastic to be true, and the findings were dismissed as the result of accidental contamination. Based on the discovery of organics in the Orgueil meteorite, Robert Robinson (1886–1975, winner of the Nobel Prize in chemistry in 1947 for his synthesis of tropinone, a precursor of cocaine.) of the

University of Oxford, suggested in 1964 that "life on Earth has evolved from organisms that were older than the Earth as we know it". He went on to say that "... our biochemistry extends beyond the Earth. In fact, there may well be one biochemistry, just as we are already well assured there is one chemistry, throughout the Universe". This was quite a bold statement, considering the amount of evidence that it was based on. Obviously for this hypothesis to be taken seriously, one has to look for more examples of organics in meteorites.

On September 28, 1969, a carbonaceous chondrite fell near the town of Murchison in Australia and many fragments were quickly recovered. Amino acids were quickly identified in the Murchison Meteorite (Fig. 2.8), opening up the study of extraterrestrial organic matter and the beginning of a gold rush looking for other organic compounds in meteorites. Analysis has shown that the amino acids in the Murchison meteorite contain both the right-handed and left-handed varieties. Since only the right-handed kind is present in living things, these amino acids are not the result of human contamination, but are intrinsically present in the meteorite and are truly extraterrestrial in origin.

Since that time, scientists have learned that these extraterrestrial rocks contain a large variety of pre-biotic organic matter. By the end of the twentieth century, over 650 individual organic molecules had been identified in primitive meteorites. These identifications are achieved by melting pieces of the meteorites using different solvents, then extracting and analyzing various components of the soluble fraction. Among the extractable organic matter were organic molecules such as aliphatic hydrocarbons, amines, amides, alcohols, aldehydes, ketones, and also biologically important compounds such as sugars, carboxylic acids, amino acids, and aromatic hydrocarbons. Most significantly, scientists found nucleobases that make up DNA and RNA (adenine, guanine, uracil, thymine, cytosine, etc.); and other important biochemical molecules (hypoxanthine, xanthine, etc.).

Since the nucleobases (Fig. 15.1) are used to encode genetic information in nuclei acids (DNA and RNA), the detection of extraterrestrial nucleobases is of great interest. However, their detection in meteorites is particularly difficult as their abundance is low and there is always the chance of contamination during the extraction and analysis procedures. From the analysis of Murchison and the Lonewolf Nunataks 94102 meteorites, Michael Callahan of the Goddard Space Flight Center in Maryland and his co-workers were able to discover nucleobase analogs which are rare in the terrestrial environment. It is therefore unlikely that these pre-biotic molecules are the result of terrestrial contamination, thus providing concrete evidence that ingredients of life were available on the primordial Earth. Furthermore, this suggests that alternative nucleobases and genetic materials could be constructed out of these prebiotic molecules, possibly leading to a new system of life. If this is true, then life in the Universe could take many diverse forms.

This list of biomolecules grew tremendously when Philippe Schmitt-Kopplin and his group at the Helmholtz-Zentrum Muenchen-German Research Center for Environmental Health, Institute for Ecological Chemistry in 2010 identified 140,000 molecular compounds with millions of diverse structures in the soluble

purine   adenine   2,6-diaminopurine   6,8-diaminopurine

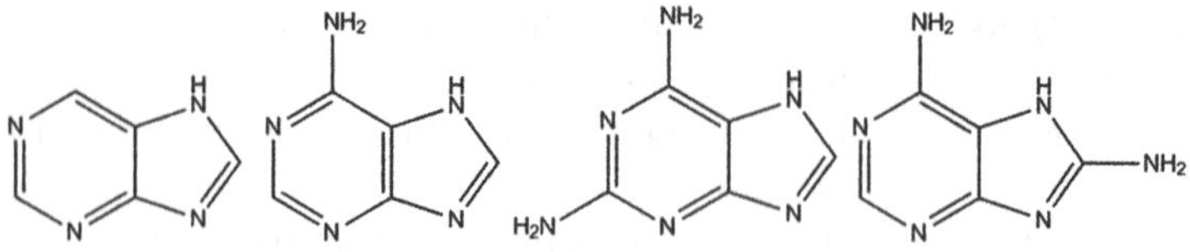

Fig. 15.1 Nucleobases found in meteorites.
Purine, adenine, 2,6-diaminopurine, and 6,8-diaminopurine are some of the nucleobases found in meteorites. These prebiotic molecules can be used to construct genetic molecules that are different from our own. Each unmarked corner of these schematics has one carbon atom

component of the Murchison meteorite. It is fair to say that almost all biologically relevant organic compounds are present in carbonaceous meteorites, although our assumption is that these are all of non-biological origin.

Interestingly, a large part of these primitive meteorites remains insoluble. This insoluble component also turns out to be made of organics, but a more complex kind. It consists of large networks of organics which chemists label as macromolecular solids. In the meteorite community, it is commonly referred to as Insoluble Organic Matter (IOM). It is estimated that the IOM represents the majority (>70 %) of the organic matter in these meteorites.

Because of the insolubility, it is not easy to determine the chemical composition of IOM. If we wish to preserve the sample, the only way is to examine it with microscopes (yielding a picture), and to perform spectroscopic measurements of the solid (yielding a spectrum). If one is willing to sacrifice part of the sample, one can heat up the sample in order to observe the gaseous products released in the process. This process is analogous to our everyday experience of cooking. The gaseous components released can be separated by techniques of chromatography, with their compositions identified by mass spectroscopy.

A more extreme procedure is burning the sample, which involves chemical reaction with oxygen. Since the chemical composition is altered, it may be more difficult to reconstruct the original composition from the products.

Through the 1980s, the work by John Cronin and Sandra Pizzarello of the Arizona State University, as well as groups in other parts of the world, established that the chemical composition of IOM is mainly made up of the elements of carbon (C), hydrogen (H), oxygen (O), nitrogen (N). These four elements are collectively called CHON by planetary scientists. In addition, there are minor amount of sulfur (S) and phosphorus (P). Most importantly, it was established the IOM is primarily organic in content. It did not escape the notice of scientists that the elements C, H, O, N, S, and P, are precisely those which are responsible for life on Earth. Carbon and hydrogen form the basis of all the biomolecules. Oxygen, together with hydrogen, forms water. Nitrogen is an essential constituent of DNA, RNA, and proteins.

Modern pyrolysis gas chromatography (in simple terms, cooking followed by separation) characterizes IOM as predominantly aromatic, with solid-state nuclear magnetic resonance spectroscopy finding various functional groups. Release of oxygen-containing molecules such as phenols, propanone, and nitrogen heterocyclics after pyrolysis suggests the IOM contains impurities beyond the basic elements of carbon and hydrogen. The structure of IOM can therefore be summarized as a complex organic solid composed of aromatic and aliphatic functional groups, as well as oxygen-containing functional groups. In 2010, Sylvie Derenne and François Robert of the Museum of Natural History of France analyzed the Murchison IOM with a variety of techniques and proposed a molecular structure that has small aromatic units but extensive aliphatic branches.

Comparison of the IOM in the Murchison, Tagish Lake, Orgueil meteorites, and the EET92042 meteorite recovered in the Elephant Moraine region of Antarctica shows remarkable similarity. George Cody and his group at the Geophysical Laboratory of Carnegie Institution of Washington performed extensive laboratory analysis of the Murchison meteorite and found that the IOM is composed primarily of highly substituted single ring aromatics, substitute furan/pyran moieties, highly branched oxygenated aliphatics, and carbonyl groups. These results were first presented by George Cody at the International Astronomical Union symposium on Organic Matter in Space in Hong Kong in 2008, and later published in the *Proceedings of the National Academy of Sciences* in 2011. Roughly speaking, the chemical structure of IOM resembles that of kerogen on Earth.

Remarkably, similar mixed aromatic/aliphatic structures are also found in organic matter in interplanetary dust particles (Chap. 2). George Flynn of the State University of New York at Plattsburgh analyzed the properties of interplanetary dust particles using a variety of techniques. From both X-ray absorption near-edge structure spectroscopy and infrared spectroscopy studies, he found that interplanetary dust particles contain a significant amount of aliphatic hydrocarbons.

The first hint that organic matter was present in asteroids came from their red color. Asteroids radiate by reflecting sunlight, and the extent of their ability to reflect indicates their chemical makeup. The extreme red colors and low reflectivity of some asteroids are difficult to explain by minerals or ices, but can be accounted for by polymer-type organic compounds which are structurally similar to kerogen. Dale Cruikshank of NASA Ames Research Center was able to fit the optical properties of the extremely red Centaur object 5145 Pholus by diffusely scattered sunlight from tholins and therefore indirectly inferred that tholins were a major constituent of asteroids.

In several chapters of this book, we have emphasized the role of carbon as an essential element in organic molecules and solids, and the role that old stars play in the nucleosynthesis of the element carbon. However, life as we know it would not be possible without another element: nitrogen. The bases of DNA molecules that carry genetic information—adenine (A), guanine (G), thymine (T) and cytosine (C)—all contain nitrogen atoms substituting for carbon atoms in their rings. Amino acids, the basic structural units of proteins, all contain an amino group with the nitrogen atom as the pillar. If extraterrestrial organics have any role in the origin of life on Earth, they must include nitrogen as a component.

In 1979, the astronomer, Carl Sagan, well known around the world for his book and TV series *Cosmos*, was performing a variation of the Miller-Urey experiment in his laboratory at Cornell University by simulating conditions in the atmospheres of planetary satellites such as Titan and Triton. By mixing some of the gases (such as nitrogen, ammonia, and methane) believed to be commonly present in the Solar System and subjecting them to ultraviolet light and electric discharge, Sagan and his associate Bishun Khare found a strange, brownish substance deposited on the flask. Since this substance did not correspond to any known terrestrial material, Sagan called it "tholins", after the Greek word for "muddy". Chemical analysis of tholins showed they have a disorganized structure, unlike minerals which have neatly repeatable arrangements of atoms. In fact, tholins are not specific, well-defined chemical compounds, but refer to a class of artificially created organic compounds made of hydrogen, carbon, and nitrogen atoms.

Curiously, at the same time at Cornell University, Jonathan Gradie and Joseph Veverka noticed that a certain class of asteroids had very red colors, and their optical properties were most consistent with the color reflected off complex organic solids. This represented a radical departure of ideas prevalent at the time because most planetary scientists believed that Solar System bodies were made of ices or minerals (see Chap. 2). Dale Cruikshank, a planetary scientist working at NASA Ames Research Center near Mountain View, California, made use of the optical properties of tholins measured by Khare (who also has moved to Mountain View) to quantitatively test this idea. Assuming that tholins were covering these red asteroids, he calculated the expected reflected sunlight spectrum from these materials and compared them to the observed spectra of the objects. The surprisingly good fit convinced him and others that organic solids are common in the Solar System.

It has been known for some time that in addition to water ice, frozen hydrocarbons are also present in the icy surfaces of many planetary satellites, including Triton, Ganymede, and Callisto. It would therefore be desirable to explore the satellites of planets. The *Cassini mission* was a joint NASA/ESA mission designed to study the planet Saturn and its satellites. The spacecraft was launched on October 15, 1997. In order to make the long journey to Saturn, the spacecraft carried 33 kg of plutonium-238 and relied on the natural nuclear decay of this radioactive element for power. Even with this nuclear engine, it still had to rely on the gravitational pull of the inner planets. *Cassini* had two flybys of Venus in 1998 and 1999, one flyby of the Earth in 1999, one flyby of the asteroid 2685 Masursky in 2000, and a flyby of Jupiter in 2000 (Fig. 15.2). After several years of interplanetary voyage, it arrived at Saturn and entered its orbit on July 1, 2004. Spectroscopic observations with the Visible-Infrared Mapping Spectrometer (VIMS) on board the *Cassini* spacecraft of Saturnian satellites, Iapetus, Phoebe, and Hyperion, also revealed evidence for aromatic and aliphatic hydrocarbons.

Onboard the *Cassini* spacecraft was a *Huygens probe* which was released from the spacecraft on January 14, 2005 and descended to the surface of Titan 2½ h later. Observations were made of the Titan atmosphere by the probe during its descent and of the surface after landing. *Cassini-Huygens* observations of Titan have suggested that methane is being converted to complex hydrocarbon-nitrile compounds in the atmosphere of Titan. The condensation of these nanoparticles (or macromolecules) on the surface resulted in dunes and lakes found by *Cassini* RADAR observations. The amount of carbon in the form of methane in the Titan

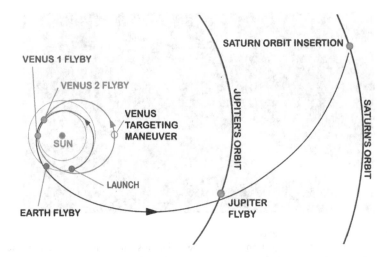

Fig. 15.2 Trajectory of *Cassini* spacecraft.
In order to reach the planet Saturn, the *Cassini* spacecraft needed to use the gravitational energies of the planets Venus, Earth, and Jupiter to give it the required propulsion. The spacecraft journey requires sophisticated dynamical calculations and careful timing

atmosphere is estimated to be 360 trillion tons, whereas the quantity of carbon in liquid form (ethane and methane) in lakes ranges from 16 to 160 trillion tons. Even larger is the amount of carbon in sand dunes, with an estimated inventory of 160–640 trillion tons. The total amount of hydrocarbons on Titan is estimated to be larger than the oil and gas reserves on Earth (see Table 14.1).

Most space scientists subscribed to the theory that the organic solids on Titan are made of tholins. An alternative theory involving hydrogen cyanide was vigorously promoted by Clifford Matthews of the University of Illinois at Chicago for many years. He showed that hydrogen cyanide (HCN) molecules can spontaneously group together to form hydrogen cyanide polymers. In the laboratory, HCN molecules can convert to a substance whose color gradually changes from yellow to orange to brown to black over a few days. Structurally, this polymer is amorphous and is not dissimilar to that of tholins. When this black polymer is stirred with water, amino acids and nucleobases miraculously appear, together with some large (macro) molecules.

Clifford Matthews has an interesting life story. When he was a young science student at the University of Hong Kong in 1941, the Japanese invaded the British Colony. Since he volunteered in the Hong Kong Volunteer Defense Corps, he was called up to defend the territory. After 18 days of fighting, the British surrendered to the Japanese on Christmas Day and Matthews became a prisoner of war. For about a year, he worked with other prisoners of war to build a new runway at Kai Tak Airport. After that, he was sent to Japan as a laborer with other British prisoners of war from Java and Singapore to work in a shipyard until Japan surrendered in August 1945. After the war, he studied in England and was inspired by John Desmond Bernal to pursue a life-long interest in the origin of life. He later

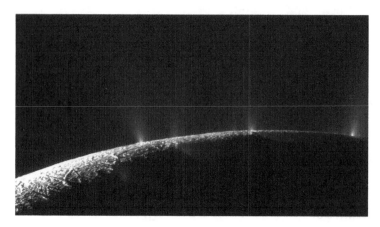

Fig. 15.3   Plumes from Enceladus.
Giant plumes are observed to eject from the surface of Saturn's moon Enceladus. This picture was
taken by the *Cassini* spacecraft. Image credit: NASA/JPL/SSI

obtained a Ph.D. from Yale and taught at the University of Illinois in Chicago until
retirement. In 1997, he returned to Hong Kong as a guest of the Hong Kong
government to witness the ceremony of handing over of the colony from British
to Chinese rule.

Since the HCN molecule is produced by old stars in large quantities (see
Chap. 9), Matthews believes they are the key to the origin of life. The best evidence
so far has been in the *Cassini-Huygens probe* to Titan, as the organics in the satellite
are likely to have contained nitrogen. Whether they are tholins or HCN polymers
remains to be determined. What is not in doubt is that organics of totally unexpected
degree of complexity are widely present in the Solar System.

Another discovery by the *Cassini mission* was the observation of the plumes
emanating from the south polar region of Enceladus, a satellite of Saturn. Enceladus
was discovered by William Herschel on August 28, 1789. Covered in water ice,
Enceladus has an extremely reflective surface, reflecting almost all the sunlight that
falls on it. On July 14, 2005, *Cassini* flew past Enceladus within 168 km and
discovered that the south polar region of Enceladus is geologically active. Large
streams flow from the surface in a form similar to the geysers seen in Yellowstone
National Park. Because of the lower surface gravity of the satellite, the plumes are
able to reach thousands of kilometers from the surface of the south pole (Fig. 15.3).
Using the Ion and Neutral Mass Spectrometer on board, the plumes were found to
be composed mainly of water, mixed with carbon dioxide, methane, acetylene, and
propane. These plumes probably originate from a liquid ocean under the icy crust.
The presence of organics in the plume raises questions on whether these seas can
harbor life. Since life forms in subsurface oceans would be extremely difficult to
detect even if they exist, the plumes of Enceladus offer an opportunity to sample the
composition of the subsurface oceans.

In this chapter, we have learned that many Solar System bodies have naturally
present organic compounds within them. These objects range from the microscopic

interplanetary dust particles, to meteorites, to asteroids, and to satellites of planets. These discoveries were made through remote spectroscopic observations, ground-based telescopes or in situ observations by space probes. They are also found in samples collected and analyzed in the terrestrial laboratory. There is absolutely no doubt that the Earth does not have the sole possession of complex organic compounds. These organics could be made in the Solar System or they could be inherited from the primordial solar nebula. The latter possibility raises a serious question: is it possible that there are also primordial organics on Earth?

A brief summary of this chapter
Organic molecules and solids are also present in different parts of our own Solar System.

Key words and concepts in this chapter
- Many different forms of organics, including amino acids and prebiotic molecules, found in the soluble component of meteorites
- Complex, disorganized organics in the form of insoluble organic matter in meteorites
- Organics in interplanetary dust particles
- The *Cassini mission* and the discovery of large reserve of hydrocarbons on Titan
- Tholins and HCN polymers as possible forms of Titan's hydrocarbons
- Plumes from Enceladus suggest the presence of an underground organic ocean

Questions to think about
1. Titan is full of organics. Do you think there is life on Titan? Is there necessarily a connection between the presence of organics and life?
2. One type of meteorites, the carbonaceous chondrites, is rich in complex organics. What is the implication of the statement "almost all biologically relevant organic compounds are present in carbonaceous meteorites"? Where do you think these organics come from?
3. Do you agree with Robinson's statement that "there is one biochemistry" that is universal throughout the Universe? Is it possible that there is one biochemistry for Earth and another biochemistry for extraterrestrial life?
4. On Earth, organic compounds are closely tied to life. Are the organics in the Solar System related to life?
5. Are the organics in meteorites, in interplanetary dust particles, and on planetary satellites related, or do they all have separate origins?
6. The detection of alternative nucleobases in meteorites suggests that our system of life may not be unique. Can you imagine what other systems of life could exist, based on a new set of genetic materials?

(continued)

7. Are you surprised to learn of the large amount of hydrocarbons in Titan? Do you think that similar large reservoirs of hydrocarbons could exist in other planetary satellites?

8. Since the Solar System is full of organics, is it possible that the Earth also contained some amount of organics when it was formed 4.6 billion years ago? If so, how can we distinguish these primordial organics from the organics that have been synthesized since then?

Chapter 16
Stardust in Our Hands

Although every atom in our body has at one time been inside a star or was made by a star, atoms are elementary constituents of our body. They have been recycled many times through various chemical reactions, and absorbed by our bodies through eating and breathing. It seems rather unlikely that we can pick up a piece of material that was made by a star and is preserved in its original form. It was therefore a total shock when, in 1987, Ernst Zinner and his group at Washington University in St. Louis, Missouri, were able to identify certain diamond and silicon carbide particles in primitive meteorites to be of pre-solar origin based on their isotopic composition. Isotopes are the same chemical elements but with a different number of neutrons in their nuclei. For example, the element oxygen can have 8, 9, or 10 neutrons in addition to 8 protons in the nuclei. Usually, the ratios among different isotopes of the same elements are determined by their nucleosynthesis history, and objects in the Solar System all have pretty much the same isotopic ratios. Stellar materials, however, have different isotopic ratios than the Solar System and therefore can be distinguished. These small solid grains can be traced directly to carbon stars or supernovae, giving the first direct evidence that stardust can be transported intact across the Galaxy to the Solar System.

Finding stardust in meteorites is not an easy task. A larger silicon carbide grain will have sizes of less than a micrometer. A typical pre-solar diamond grain has a size of a few nanometers and contains only about 1,000 atoms. In order to identify the minuscule pre-solar grains preserved in the interior of a meteorite, one has to melt away the outer parts. In the words of Edward Anders of the University of Chicago, it is like "burning down the haystack to get at the needle". With modern electron microscopes, we can actually take pictures of these stellar grains, and two examples of SiC and graphite pre-solar grains are shown in Figs. 16.1 and 16.2.

As with any great discovery, this piece of news was received with skepticism by the scientific community. However, with time, other pre-solar grains including graphite, silicates (olivine, pyroxene), corundum, and spinel were identified. The fact that grains of similar chemical composition have been found in stellar atmospheres through infrared spectroscopy (Chap. 11) has raised the possibility

S. Kwok, *Stardust*, Astronomers' Universe, DOI 10.1007/978-3-642-32802-2_16,
© Springer-Verlag Berlin Heidelberg 2013

Fig. 16.1 Electron
microscope image of a pre-
solar SiC grain.
The size of the grain can be
seen in comparison to a bar of
length 1 μm shown in the
lower left. Image credit:
E. Zinner

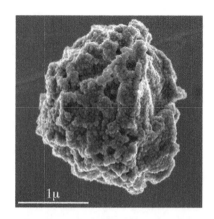

Fig. 16.2 Electron
microscope image of a pre-
solar graphite grain.
Image credit: E. Zinner

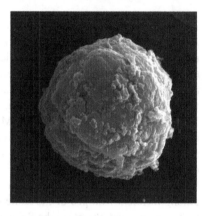

that the rocks we hold in our hands contain micrometer-size pieces of matter
delivered to us from old stars.

In October 1996, an international conference was held on the campus of
Washington University in St. Louis, Missouri, to celebrate the tenth anniversary
of the discovery of stardust on Earth. For the first time, astronomers were brought
together in the same room with scientists who study meteorites. While Ernst Zinner
and his group were able to determine the isotopic ratios of elements in pre-solar
grains with great accuracy, they had to rely on astronomers such as John Lattanzio
of Monash University in Australia, who calculates expected isotopic ratios of
elements in AGB stars as the result of stellar nucleosynthesis, to tie the origin of
the pre-solar grains to stars.

Although the two groups hardly knew each other before the meeting and often
spoke very different technical languages, this gathering at St. Louis was the
beginning of a new era of stellar mineralogy. We used to think that astronomy
was a purely observational science. Because stars are so far away, we can only
watch them but not touch them. With stardust in meteorites, astronomy has taken on
an experimental component, where distant celestial materials can be put in a lab
instrument and be subjected to detailed analysis.

Fig. 16.3 The NASA U-2 aircraft.
NASA has converted this high-flying reconnaissance aircraft into a flying laboratory to collect dust
particles in the upper atmosphere

Fig. 16.4 Samples of
interplanetary dust collected
by the NASA U2 aircraft.

Now, instead of waiting for meteorites containing pre-solar grains to fall onto
Earth, what if we actually go out and capture them? We already know that the Solar
System contains a large amount of interplanetary dust particles (Chap. 2), and all
we would have to do is collect some and bring them back to Earth. This
sounds simple in principle, but is not so easy to carry out. In the 1990s, Donald
E. Brownlee of the University of Washington flew the NASA *ER2* aircraft
(Fig. 16.3) to the stratosphere at 20–25 km altitude to collect interplanetary dust
particles. The *ER2* (abbreviation for *Earth Resources-2*) aircraft is the civilian
version of U.S. Air Force's U-2 reconnaissance plane. The Lockheed built high-
flying aircraft was introduced in 1957 to perform spy missions over the Soviet
Union. Not only could it fly above the altitudes of ground-to-air missiles and
intercepting fighters, it was also outside the range of radar at that time. The airplane
and the secret flights only came into the public eye in 1960 when an U-2 plane was
shot down in the Soviet Union and the pilot, Francis Gary Powers, was captured
alive. In 1971, NASA's High Altitude Aircraft Program acquired two U-2Cs from
the U.S. Air Force and began using the aircraft for scientific purposes at the Ames

Research Center. Flying at the edge of space, the *ER2* has performed many functions in remote sensing and Earth observations. Of particular significance is its program to sample particles in the stratosphere.

Although interplanetary dust particles (Fig 16.4) collected by high-flying aircrafts could be contaminated by terrestrial particles coming into the stratosphere from volcanic eruptions, above-ground nuclear weapons tests, and exhaust from rocket fuel, they can be distinguished from these contaminants by their chemical compositions. Interplanetary dust particles are found to consist of a mix of silicates, oxides, and various carbonaceous materials such as amorphous carbon, nanodiamonds, graphite, and most interestingly, macromolecular organics (Chap. 15). Isotopic analysis of interplanetary dust particles shows evidence of deuterium enrichment, with deuterium to hydrogen ratio as high as 50 times the value normal in the Solar System. This suggests that at least some of these particles have arrived from outside of the Solar System.

It is now clear that old red giant stars can make a variety of minerals, including rare gems like diamonds and sapphires (Chap. 11). They do so under the most unfavorable conditions, at extremely low densities and very high temperatures. Furthermore, the synthesis takes place on such a large scale that the stars can spread these materials across the Galaxy. They travel across the Galaxy and mix with interstellar clouds from which the solar nebula was born. Most amazingly, we now can hold in our hands (in the form of meteorites) the materials that we know are made in stars. This stardust may have taken a long time to reach us, but there is no doubt about its celestial origin. It may be another century or two before humans travel to stars, but it is comforting to know that every year stars deliver presents to us in the form of gems.

The Earth may seem immense to us, but on the Galactic scale, the Earth is but a speck of dust. On the whole, 48 % of all atoms of the planet Earth are oxygen, 15 % are iron, magnesium 16 %, and silicon 15 %. These four elements therefore represent 95 % of all atoms on Earth.[1] These ratios are very similar to those of silicate dust found in oxygen-rich red giant stars. Carbon, the fourth most common element in the Universe, is under-represented on Earth. Carbon is found in carbonate rocks, in the Earth's atmosphere (mostly in the form of carbon dioxide and a small amount as methane[2]), in the oceans, and as part of living or dead organisms (Table 14.1). Carbon comes in 12th in the fraction of atoms (1.6 %) on Earth.

If minerals made in stars have reached the Solar System, how about organics? This important question was pursued by many scientists, who searched for isotopic evidence in organic matter in meteorites and interplanetary dust that these materials also carry non-Solar System signatures. The similarity between the 3.4-μm aliphatic

---

[1] These fractions are expressed in relative number of atoms. If expressed in mass, then the largest fraction in mass is in iron because it is heavier than oxygen. The mass fractions of Earth are iron (32 %), oxygen (30 %), silicon (16 %), magnesium (15 %) (Appendix E).

[2] Both atmospheric carbon dioxide and methane are of high interest today because of their roles as greenhouse gases. The contribution of men-made carbon dioxide is believed to be a major cause of global warming.

feature observed in meteorites and the absorption feature seen in the Galactic Center (Chap. 17) led Pascale Ehrenfreund of Leiden University and Yvonne Pendleton of NASA Ames Research Center to separately suggest that the diffuse interstellar space contains organic material similar to material found in the IOM (Chap. 15). Excesses of deuterium (D, or ^{2}H), carbon 13 (^{13}C) and nitrogen 15 (^{15}N) isotopes in these materials suggest possible interstellar origin.[3]

Luann Becker of the University of Hawaii showed in a paper published in the *Proceedings of the National Academy of Sciences* in 2000 that the anomalous isotopic ratios of noble gases trapped in fullerenes C_{60} to C_{400} in the Allende and Murchison meteorites point to an extraterrestrial origin. The famous Tagish Lake meteorite that we first talked about in Chap. 2 shows elevated ratios of ^{15}N/^{14}N and D/H in its organic globules, leading Keiko Nakamura-Messenger of the NASA Johnson Space Center to suggest that interstellar organics exist in the present Solar System.

In the conference on *Cosmic Cataclysms and Life* held at the European Space Agency in Frascati, Italy, in 2008, Pascale Ehrenfreund of the Astrobiology Laboratory of Leiden University reported that she and her student, Zita Martins, had discovered amino acids possessing isotopic signatures consistent with interstellar origins in two ancient meteorites found in Antarctica. They also found that the carbon isotope ratios of nucleobases in the Murchison meteorite show signs of being interstellar in origin. The implications of these discoveries are clearly spelled out in a bold statement they wrote in their paper in *Earth and Planetary Science Letters*: "These new results demonstrate that organic compounds, which are components of the genetic code in modern biochemistry, were already present in the early Solar System and may have played a key role in life's origin".

Although early theories on the origin of meteoritic organics concentrated on synthesis in the solar nebula, the latest evidence suggests that the solar nebula only played a role of secondary processing of pre-existing interstellar material. There is increasing appreciation that the early solar nebula might have inherited a significant amount of interstellar organics, and these organics survived the subsequent condensation and mixing of the solar nebula. It is likely that some of these primordial organics were embedded in comets and meteorites that formed during the early days of the Solar System.

The fact that we can actually hold solid particles made by stars in our hands is truly remarkable. We have to appreciate that the stardust particles have traveled a very long way, hundreds or thousands of light years, across the Galaxy. They had to survive a variety of hazards along the way, including damage from ultraviolet light from stars, and high-velocity shocks in the turbulent interstellar space. In fact, it was a common belief among many astronomers that dust produced by stars would

---

[3] The symbol ^{13}C and ^{15}N represent carbon of atomic weight 13 and nitrogen of atomic weight 15, respectively. The carbon atomic nucleus has 6 protons and usually 6 neutrons (^{12}C). The isotope ^{13}C has one extra neutron, giving a total atomic weight of 13 (6 + 7). Similarly, the isotope ^{15}N of nitrogen has 7 protons and 8 neutrons.

be destroyed quickly once it entered interstellar space. But stardust not only survived the journey to our Solar System, it survived the collapse of the solar nebula and formation of the planetary system. Lastly, stardust survived the passage through our Earth's atmosphere, which serves as a protective shield and vaporizes a large part of solid bodies trying to enter the Earth. The fact that we do find stardust at all is a testament of its sturdiness and durability, but also a reflection of the common occurrence of stardust. Stardust is not just an optical phenomenon that we detect through a spectrograph, but is something solid, physical, and as real as any scientist can hope for. The discovery of grains of pre-solar origin in meteorites lends strong support to astronomical spectroscopic observations, giving us confidence that we have been on the right track, no matter how fanciful the idea may have first seemed.

A brief summary of this chapter

There is evidence that solid particles made in stars have traveled across the Galaxy and reached our early Solar System. The fact that we can hold stardust embedded in meteorites in our hands is the most direct evidence of the stellar-Earth connection.

Key words and concepts in this chapter

- Stardust and pre-solar grains
- Pre-solar diamonds and silicon carbide in meteorites
- Collection of interplanetary dust particles from the stratosphere
- Organics of interstellar origin
- Extraterrestrial fullerenes

Questions to think about

1. Truth may be more fantastic than fiction and "stardust in our hands" can be considered as such an example. What do you think are the philosophical implications of this statement?
2. The Earth is made up of mostly oxygen, iron, silicon, and magnesium, which is very different from the chemical composition of stars, or even other planets (such as Jupiter). Why is the Earth different?
3. According to Ehrenfreund and Martins, components of the genetic code were already present in the early Solar System. What is the implication of this statement?

Chapter 17
Bacteria in Space?

In Chap. 13 we talked about the possibility that the widely observed extended red emission (ERE) in the Galaxy could be due to nanodiamonds. This is not entirely surprising, as diamonds are made up of pure carbon, the fourth most common element in the Universe after hydrogen, helium, and oxygen. We have to remember that life as we know it is based on carbon. Is it possible that ERE comes from something more complicated than we have been assuming? It is known that many biological pigments absorb blue light and fluoresce in the red. Chlorophyll, commonly found in plants, and a large class of bacteria fluoresce in red light. This hypothesis was put forward by Fred Hoyle and Chandra Wickramasinghe in 1996. When this was first proposed, the authors were ridiculed by the scientific establishment, and some even believe that this cost Fred Hoyle the Nobel Prize as the Prize for stellar nucleosynthesis was only awarded to Fowler and not to him (Chap. 4). Is it possible that there are biological organisms in space, and how can we test this hypothesis?

The biomolecules in our body can be broadly classified into four groups: proteins, nucleic acids, lipids and carbohydrates. Sugar, found naturally in fruit and honey, is the simplest form of carbohydrates (meaning compounds of carbon and water). Complex carbohydrates (called polysaccharides) are made up of thousands of units of simple sugar (called monosaccharides) linked together. One example of polysaccharides is cellulose, the main constituent of plant cells. Cellulose makes up about 50 % of wood and 90 % of cotton, and represents one of the most abundant organic compounds in the biosphere.

The most important aspect of carbohydrates lies in its use as food, our energy source. Unlike sugar, cellulose cannot be consumed as food by people, but cows, sheep and horses can digest it with microorganisms in their alimentary tract. The conversion of cellulose into protein by these animals allows us to indirectly utilize cellulose as energy source by eating beef or lamb.

In the past 20 years, astronomers have already made significant progress in the search for biomolecules in space. As of 2012, over 160 molecules have been discovered in space, the majority of them organic molecules. The rich variety of organics in space is amazing: the list includes hydrocarbons (e.g., methane CH_4,

S. Kwok, *Stardust*, Astronomers' Universe, DOI 10.1007/978-3-642-32802-2_17,
© Springer-Verlag Berlin Heidelberg 2013

Fig. 17.1 The 100-m *Robert C. Byrd Telescope* in Green Bank, West Virginia.
The telescope is built in this remote valley to avoid man-made radio interference. It is one of the largest movable structures on land. Since its completion in 2000, the *Green Bank Telescope* has been used for different kinds of astronomical research including the search for complex molecules in space. Photo credit: NRAO

acetylene C_2H_2, ethylene C_2H_4), alcohols (e.g., methanol CH_3OH, ethanol C_2H_5OH, vinyl alcohol $H_2C=CHOH$), acids (e.g., formic acid $HCOOH$, acetic acid CH_3COOH), aldehydes (e.g., formaldehyde H_2CO, acetaldehyde CH_3CHO, propenal $CH_2=CHCHO$, propanal CH_3CH_2CHO), ketones (e.g., ethenone $H_2C=CO$, acetone, CH_3COCH_3), amines (e.g., methylamine CH_3NH_2, cyanamide NH_2CN, formamide NH_2CHO), ethers (e.g., dimethyl ether CH_3OCH_3, ethyl methyl ether $CH_3OC_2H_5$), etc. Among the detected organic molecules are some very large ones, such as ethylene glycol ($HOCH_2CH_2OH$), propenal (CH_2CHCHO) and propanal (CH_3CH_2CHO), acetone (CH_3COCH_3), cyanoallene (CH_2CCHCN), acetamide (CH_3CONH_2), cyanoformaldehyde ($CNCHO$), and cyclopropenone ($c–H_2C_3O$). These detections were made possible by the construction of new, large telescopes such as the 100-m *Green Bank Telescope* (Fig. 17.1).

Stars and interstellar clouds are very far away from Earth. How can astronomers be so sure about their molecular identification? In fact the requirement for detection is very stringent. The agreement between the astronomically measured frequency for the spectral line has to be within one in ten millionth of the laboratory measured molecular frequency. This is possible as modern spectrometers have very high spectral resolutions. The concept of the spectral resolution of radio telescopes can be illustrated by the tunings of radios. In the order to search for a particular radio station, we have to scan the radio band to seek out the signal from a station broadcasting at a particular frequency. A radio receiver with fine tuning ability will be able to separate a station broadcasting at one frequency from another broadcasting at a nearby frequency, thereby avoiding the mixing of signals. Since there are many different molecules each broadcasting at their own frequencies in space, we need a very fine tuning receiver to separate all these signals.

In order for an interstellar molecule to be positively identified, multiple transitions from that molecule must be observed to confirm a molecular identification. For more complex molecules, more transitions are required. Not only are multiple lines needed, their strengths must also have the correct ratios. Although it is not easy to convince laymen of this seemingly difficult task, the identification of interstellar molecules is in fact very robust.

Of particular interest are prebiotic molecules leading to the formation of proteins, carbohydrates, nucleic acids, and lipids. Using the *National Radio Astronomy Observatory* 12 m telescope (Fig. 9.2) at Kitt Peak, Arizona in 2000, Jan Hollis of the NASA Goddard Space Flight Center, Frank Lovas of the University of Illinois, and Phil Jewell of the *National Radio Astronomy Observatory*, were able to detect glycoladehyde, the first sugar in space. Active searches are now underway for the simplest amino acid glycine (NH_2CH_2COOH) and for the parents of the bases that constitute the structural units of DNA and RNA, such as purine (c–$C_5H_4N_4$) and pyrimidine (c–$C_4H_4N_2$). Pyrimidine is the base contained in cytosine (DNA and RNA), thymine (DNA), and uracil (RNA), whereas purine is the base for adenine (DNA and RNA) and guanine (DNA and RNA). With increasing sensitivities of radio telescopes, it will not come as a surprise when these prebiotic molecules are discovered in the near future.

We have noted that the elements most common in living things, including oxygen, carbon and nitrogen, are also among the most cosmically abundant. But there is one exception and that is phosphorous. Phosphorous is an important element in biochemistry and is involved in energy transfer and membrane structure, for example. But the relative abundance of phosphorous in the human body is several orders of magnitude larger than in the Solar System—where it ranks just 17th in the list of the most common elements.

So, how did phosphorous concentrate on Earth to, ultimately, become part of us? This conundrum was raised by Enrique Maciá of the Complutense University of Madrid. He reviewed the synthesis of phosphorous, the forms in which it can hide itself, and measurements of its abundance. He speculated that the element preferentially condenses into solids, getting incorporated into comets that deliver it to Earth.

That provides a mechanism for the element's arrival, but leaves us wondering about the fraction of interstellar phosphorous in molecular or solid forms. Astronomers have detected many molecules containing carbon, nitrogen and oxygen in the interstellar medium and around stars. But curiously, the only known phosphorous-bearing astrophysical molecules are phosphorus mononitride (PN), carbon monophosphide (CP), and phosphorus monoxide (PO). Is it possible that phosphorous hides in a form that we fail to recognize?

Although the radio technique yielded many successes in detecting interstellar molecules, it does have some limitations. Since many molecules radiate in the radio region, a search for weak signals from rare molecules will eventually be crowded out by the loud signals from simple molecules. It is like living in a big city with many radio stations. Even with a very sensitive radio, you will have a hard time picking up ever distant stations as the broadcasts from a large number of closer stations fill up every frequency band.

Man-made interference

The modern society has created many commercial devices that make use of the radio spectrum. First we have radios and television, now we have cell phones, wireless phones, radio controlled planes, ham radios, walkie-talkies, satellite communications, etc. Even the microwave ovens in our kitchens are generating a lot of radio noise. Now with many individuals of the Earth's population owning a cellular phone and the need for data transfer increasing each day (going 3 G and 4 G), our sky is filled with radio signals.

Each of these communication and broadcasting devices has to be assigned a certain part of the radio spectrum for its use. Since the signals from these devices are infinitely stronger than the natural signals we receive from interstellar molecules, it is easy to see that all these man-made devices will eventually crowd out all the signals from the universe and make radio astronomy impractical. This is why there is an international agency to regulate the use of radio devices. The agency is trying desperately to reserve certain radio bands for astronomical use (Fig. 17.2).

There are very few corners of the world left that are isolated from man-made radio interference. Certain regions in Western Australia, western China, northern Canada, Greenland, and Antarctica are some examples. If these regions are invaded by civilization, astronomers may need to go to the Moon to get some peace and quiet.

We have talked about how astronomical infrared spectroscopy has been used to identify aromatic and aliphatic compounds in stars (Chap. 10). Could these organic compounds be part of a larger biological system? For example, polysaccharides are structural elements in the cell walls of bacteria and plants and Fred Hoyle and Chandra Wickramasinghe have suggested that the UIR features could be due to bacteria. Is it possible that bacteria are made in stars and exist in interstellar space as they transit from one stellar system to the other?

Bacteria are also the oldest form of life. The earliest fossil evidence of the existence of bacteria on Earth can be traced back 3.4 billion years. About 2.7 billion years ago, a new species of photosynthetic bacteria appeared in the oceans. These bacteria, called cyanobacteria, were the first organisms to produce oxygen and are believed to be responsible for introducing oxygen molecules to the Earth's atmosphere. Bacteria also have the uncanny ability to convert the nitrogen molecules in the atmosphere to ammonia, which makes it possible for other organisms to synthesize proteins and nucleic acids. It is not an exaggeration to say that without bacteria, most life forms would not exist on Earth today.

Bacteria, as we now know, are very hardy beings. They emerged on Earth at a time when the Earth was a very inhospitable place (from our point of view and by today's standards). The atmosphere had no ozone layer to block out the harmful ultraviolet light from the Sun. There was no oxygen to breathe for metabolism.

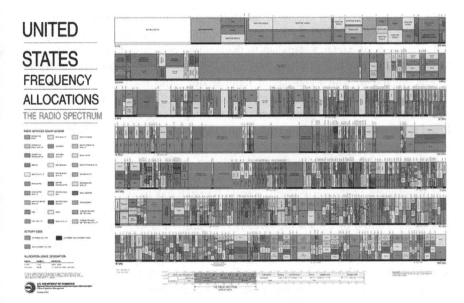

Fig. 17.2 United States radio frequency allocation chart.
This busy chart shows how the radio spectrum is being divided for commercial use. The man-made signals are much, much stronger than the weak natural signals from the Universe. The increasing utilization of radio technology has deprived astronomers of access to radio signals from space. Photo credit: US Department of Commerce

Even the radiation background released from the decay of uranium inside the Earth was much higher than it is now. A new discipline had emerged in recent years to study life forms under extreme conditions. Bacteria have been found to flourish under high temperature, high acidity, extreme dryness, and other undesirable conditions. These came as a great surprise to biologists as it had long been believed since the times of Louis Pasteur (1822–1895) that no living thing could survive in temperatures exceeding 80° C. Now it is known that life can exist at temperatures over 120° C. As late as the early 1990s, microbiologists believed that no life existed more than 7.5 m under the seabed. Now, life has been found more than 1 km under the seabed. In 2008, John Parkes of Cardiff University in the UK found off the coast of Newfoundland living microbes in rocks 1.6 km under the sea bed, which itself has 4.5 km of seawater above it.

New names have been given to creatures that survive under strange, extreme conditions. Those that can exist under high temperatures are called "thermophiles", under high acid conditions "acidophiles", in salty conditions "halophiles", under high pressures "piezophiles", in extreme dryness, e.g., in deserts, "xerophiles". Collectively, they are referred to as "extremophiles". The earliest examples of the existence of halophiles can be found in the Old Testament, where a red bloom was noticed in the Dead Sea, which was known to be highly salty. This red bloom is now believed to be due to bacteria that can live under such conditions.

In the late 1970s, Carl Woese of the University of Illinois found in his study of the genetic makeup of bacteria through DNA sequencing that there are actually two different kinds of bacteria. This led to the creation of a new domain of life, now named archaea. The tree of life as envisioned by Woese consists of three domains. One is eukaryota, which consists of familiar life forms such as plants, animals, fungi (e.g., mushrooms), and protists (e.g., amoeba). The other two domains are bacteria and archaea. Many archaeans are extremophiles. For example, archaea are found to exist in hot springs where the water temperature is above the boiling point of 100° C.

Bacteria and archaea did not exist in isolation. In spite of their small sizes, they were in fact responsible for the alteration of the chemical composition of the oceans and atmosphere of the early Earth. Recent research has found that they are a lot tougher than we thought. A colony of bacteria has been found in a layer of soft red clay at the bottom of the Pacific Ocean. They are found to have survived for 86 million years on a minimum amount of oxygen and nutrients. This implies that this community has been there since the days of the dinosaurs. If bacteria can live in such an unfavorable environment, are the conditions in space that much more hostile?

One area of research that has generated a high degree of fascination is the possible existence of a deep biosphere. Although most of our familiar life forms are located on the surface of the Earth, an increasing number of varieties of microbes are found deep under ocean floors, buried in sediments, and kept in oil reservoirs. It is quite conceivable that bacteria and archaea in the oceanic crust in fact represent the largest fraction of biomass on Earth. Traditionally, we think of life as limited to a thin layer above the solid crust of the Earth. Now it is conceivable that this layer extends deep into the interiors of the Earth. The fact that life can exist under unusual conditions means that life is versatile, and can exist under a much wider range of physical conditions. Our previous views of life may be chauvinistic.

If this is the case, life, at least simple life forms such as bacteria and archaea could quite possibly exist on Mars, on planetary satellites such as Titan and Europa, and may be even on asteroids and comets. If we really stretch our imagination, how about the possibility of life in interstellar space? Our Milky Way galaxy is filled with clouds of gas and dust (Fig. 17.3). Could life exist in such dilute environment? Even if life is not as densely clustered together as it is on Earth, these clouds are huge and as collective entities, they may contain a lot of organisms.

This is entirely speculative, of course. But this is the idea that Hoyle and Wickramasinghe wanted to pursue. Chandra Wickramasinghe has a brother, Dayal Wickramasinghe, who was a professor at the Australian National University in Canberra, who came to visit the U.K. almost every year in the 1970s. When Fred Hoyle and Chandra told Dayal about their ideas on bacteria in space, Dayal asked Fred Hoyle what was the best way to detect bacteria in space. Hoyle replied it would be to search for their infrared signatures in absorption towards the Galactic Center as the telescope would look through a very long path, and the accumulated strengths of the signal along the line of sight would be the strongest.

As we have discussed in Chap. 10, infrared spectroscopy can detect the vibration of molecules, and a complex, organic molecule can have many different ways of

Fig. 17.3 *Spitzer Space Telescope* infrared image of M8.
This infrared image of the Lagoon Nebula (M8, NGC 6523) in the constellation of Sagittarius is taken with the Infrared Array Camera of the *Spitzer Space Telescope*. The color image is a composite of images taken at 3.6 (*shown as blue*), 5.8 (*shown as green*), and 8 μm (*shown as red*). The reddish color nebulosity is due to emission from solid-state dust particles. Could these interstellar dust clouds harbor simple life forms?

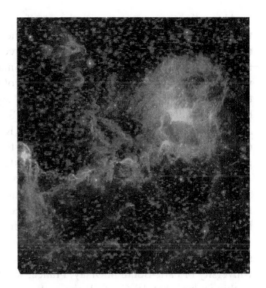

vibration (stretching, bending, etc.) that can be used as signatures of its presence. One of the most commonly observed infrared features characteristic of organic compounds is the 3.4 μm feature originating from the C–H stretch of methyl (CH_3) and methylene (CH_2) groups. The possibility of detecting organic materials in the diffuse interstellar medium through the observation of this 3.4-μm feature was first raised by Walt Duley and David Williams in their paper in *Nature* in 1979.

When Dayal Wickramasinghe returned to Australia, he applied for telescope time at the 3.9-m *Anglo Australia Telescope (AAT)* to carry out this experiment. The proposal was promptly turned down by the time allocation committee on the grounds that the proposal was too speculative. However, he was able to convince the infrared astronomer David Allen to give it a try unofficially. During one of the observing periods when the weather was considered to be not good enough for optical observations, David Allen changed the instrument to infrared and pointed the telescope at the Galactic Center. To his surprise, he detected a very strong absorption feature at the wavelength of 3.4 μm. The report of this detection was published in *Nature* in 1980.

Although the bacteria hypothesis was the motivation for the original observations by Wickramashinghe and Allen, I should point out that the detection of the 3.4 μm feature does not constitute proof of the existence of bacteria in space, as this spectral feature is common in organic compounds with aliphatic structures, for example, in the case of lipids (as in fat). More specifically, this feature corresponds to the natural frequency of methyl and methylene chemical groups undergoing stretching motions. Any organic compound with this structure will be able to absorb light, and the light energy taken in by the molecule will energize the molecular groups into stretching motions. Examples of natural substances that exhibit this feature are coal and kerogen (Chap. 10). This feature has been seen in many different directions, showing that organic compounds are widely present in the diffuse clouds between stars.

Absorption spectroscopy is a powerful technique because if we assume that organic materials are widely present in the Galaxy, then all the cold materials along the line of sight to a distant bright source will absorb the light coming from that background source. If there are sufficient materials along the line of sight, they will cause a dip (absorption) in the spectrum of that source. The 3.4 μm feature in the diffuse interstellar medium along lines of sights to more sources was detected by Scott Sandford and Jean Chiar of the NASA Ames Research Center using the *NASA Infrared Telescope Facility* on Mauna Kea in 1989, and the *Infrared Space Observatory* respectively. After the launch of the *Spitzer Space Telescope*, the 3.4-μm absorption feature was found in many external galaxies.

These results suggest that organic compounds are not just confined to the surroundings of stars but are widely distributed in the Milky Way Galaxy. The fact that this feature can be detected in external galaxies means that there must be a large quantity of organic compounds in the Universe.

These results show that organic compounds fill the volumes of space in between stars in galaxies. Emmanuel Dartois of the Institut d'Astrophysique Spatiale, University of Paris-South in Orsay, France, was able to estimate from the strengths of this feature that 15 % of all the carbon atoms in the Universe are tied up in aliphatic organic compounds. If we recall that carbon atoms can be in the form of ions (e.g., C^+, carbon atoms with one electron removed), atoms (neutral carbon atoms), molecules (e.g., CO, see Chap. 9), and solids (e.g., SiC, see Chap. 6), the 15 % fraction in aliphatic organics is a very significant number.

How did these complex organics come into being? I believe that they are built up step by step from simple molecules. Evidence for this can be found in the late stages of stellar evolution when simple molecules such as CO and HCN, are first synthesized by red giants, but as they get older, simple organic molecules such as acetylene are formed. When the stars evolve to the proto-planetary nebulae stage, we have diacetylene (C_4H_2), triacetylene (C_6H_2), culminating in the formation of the first ring molecule benzene (C_6H_6). Also in this stage of evolution, the first signs of the aromatic and aliphatic compounds appear. This sequence of events led me to believe that complex organics are built up from small to large, simple to complex, over time.

On the other hand, Chandra Wickramasinghe believes that they are due to breakdown of biological materials. We had a discussion on this issue when we met during the International Workshop on Dust, Molecules, and Chemistry at the Inter-University Center of Astronomy and Astrophysics in Pune, India, in November 2011. His argument is that life is common in the Universe. When the planets where life developed are destroyed, e.g., as the result of stellar evolution, the remnants of life are spread everywhere and the interstellar infrared signatures that we are witnessing could have their origins in life.

One of the intrinsic limitations of astronomy is that the objects we study are so far away, it is impossible for astronomers to perform in-situ experiments. It is not possible in the foreseeable future to send a space probe to a star to collect samples for analysis. The best we can do is to perform remote spectroscopic observations with telescopes and rely on the comparison between the astronomical spectroscopic data and what we know from laboratory measurements.

In an attempt to test the bacteria hypothesis of Hoyle and Wickramasinghe, Yvonne Pendleton of the NASA Ames Research Center obtained laboratory spectra of bacteria and compared them with the best astronomical infrared spectra available. She found that they do not match well and there is no convincing evidence for the existence of biological material in space.

Even if the observed spectra could be fitted by the spectra of bacteria, it would not prove that bacteria are responsible. The fact that bacteria (and many living things) are made up of the same organic molecules with aromatic and aliphatic structures implies that we cannot uniquely assign these infrared signatures to bacteria, or any other specific organism. Scientists believe in the simplest explanation for a phenomenon, a principle called Occam's Razor. If living matter is to be found in space through astronomical means, then the spectroscopic data have to be of extremely high spectral resolution so that fine distinctions between different organic matter can be made. This would require flying a very large infrared telescope in space (or placing one on the Moon) with sophisticated instrumentation not yet available.

One should always remember that the absence of evidence does not mean that we can rule out the hypothesis. For example, at present there is no evidence for extraterrestrial intelligent beings beyond the Earth, but it does not prevent many scientists from believing that intelligent beings are widely present in our Galaxy. We simply do not have the technical capability to confirm their existence.

This is also true for the bacteria hypothesis of Hoyle and Wickramasinghe. We have already shown that very complex non-biological organic compounds can be made naturally by stars and they can travel over immense distance through interstellar space to arrive in our Solar System. We have also learned from examples on Earth that simple forms of life such as bacteria and archaea can survive in very hostile environments. Gerda Horneck of the German Aerospace Center showed that bacteria, if protected inside meteorites, can survive destruction from UV radiation, low density, and cosmic rays in the interplanetary space and be transported from solar system to solar system. As fantastic as this idea may sound, what is to say that other simple life forms could not be ejected from other stellar systems that harbored life and travel through space to reach us?

While there is no evidence for the existence of bacteria in space, the interest in this possibility did lead to one significant discovery, namely observations of the wide-spread nature of the 3.4-μm emission feature in the diffuse volumes between stars and in galaxies. This feature is an indicator of aliphatic structure and its presence suggests that complex organic compounds are everywhere in the Galaxy and beyond. These organic compounds could be synthesized in interstellar space, or, more plausibly, they are made in stars and ejected into the interstellar medium. In Chap. 10, we have discussed that the 3.4-μm feature was detected in proto-planetary nebulae, giving evidence that aliphatic materials are made in stars.

No matter what the origin of the carrier of the 3.4-μm feature, is there is no doubt that there is a lot of it. The work of Dartois suggests that at least 15 % of the element carbon is contained in these organics. The implication of this statement is immense. It means that organic compounds can be made in large quantities, and be widely distributed throughout the Galaxy. If they are so wide-spread, could some of these materials end up in our Solar System? The fact that organic matter in meteorites and interplanetary dust particles also show this 3.4-μm feature has been demonstrated by Yvonne Pendleton. Does this imply a physical connection between organics in the Solar System and other stars?

In this chapter, we have shown that a large number of organic molecules and some pre-biotic molecules have been detected in the interstellar medium with radio spectroscopic techniques. The radio technique has the advantage that it can identify exactly the chemical structure of the molecule, but there are considerable technical challenges in searching for very large, very complex pre-biotic molecules. The infrared technique, on the other hand, can detect features that arise from complex organics (the 3.4-μm feature is one example), but it lacks the precision to pinpoint the exact chemical configuration of the substance. We hope that continued technical development of both techniques will gradually let us find pre-biotic molecules that make up the elements of life.

On the other front, planetary exploration will soon allow us to land instruments on Mars and other planetary satellites that will directly search for evidence of living organisms. Indeed, on August 6, 2012, the *Curiosity* rover landed on Mars. Equipped with a host of instruments, the rover is able to survey the geological structure of the Martian surface and analyze the chemical content of minerals on Mars. The identification of organic compounds and other biological signatures of life will be the first steps in the search of life on Mars to be undertaken by future Mars missions. The finding of bacteria or other simple life forms beyond the Earth would give a tremendous boost to the idea that life is common in the Universe.

A brief summary of this chapter
Prebiotic molecules are detected in space and there is evidence that there are large quantities of complex organic matter in interstellar space and in other galaxies.

Key words and concepts in this chapter
- Detection of interstellar molecules
- Alteration of the Earth's atmosphere by early life
- Extremophiles on Earth
- Deep biosphere
- Our chauvinistic view of life
- Aliphatic compounds in interstellar space
- A large fraction of the element carbon in galaxies are tied up in complex organics
- Can bacteria exist in space?

Questions to think about
1. Our Earth's atmosphere did not begin with an oxygen atmosphere. It only developed one after the beginning of life. But now we rely on the oxygen-rich atmosphere to survive. Is this a coincidence?
2. Spectroscopic techniques have proven to be a very effective way to remotely identify different kinds of matter in astronomical objects. Do you think there are any limitations to this technique?
3. In this book, we have talked about the Earth's atmosphere in different contexts. Can you speculate on the various effects of a (1) thicker/thinner atmosphere; and (2) atmosphere of different chemical composition?
4. Organic compounds of great complexity have now been discovered in space. Do you think these substances could represent breakdown products of life?

Chapter 18
Comets: Messengers from the Past

Because of their spectacular appearance, comets have commanded a strong interest among men since ancient times. The earliest record of a comet is believed to have been kept in oracle bones, which are dated to the Shang Dynasty in China over 3,000 years ago. The *Spring and Autumn Annuals*, the ancient Chinese historical text covering the period from 722 BC to 481 BC, reported the sightings of three comets, including possibly Halley's Comet: "In the fourth year of Wen Kung, Prince of Lu, (611 BC) in the autumn the seventh month, there was a comet which swept in the Pei Tow (The Great Dipper)". Through the appointment of court astronomers, the Chinese kept an impeccable record of comet sightings for over 2,000 years.

Unlike the movement of the Sun, the Moon, planets, and stars, comets appear on an irregular basis and were generally treated as omens for good or bad things to come (Fig. 18.1). An alternative to the prevailing opinion that comets were astrological signs carrying messages about the future was that comets were an atmospheric phenomenon and therefore local to the Earth, a view that was endorsed by Aristotle. Aristotle regarded comets, meteors, and the aurora borealis as all transient phenomena in the upper atmosphere, things that burst into flame and visible brightness. In the Aristotelian view, the heavens consist of the Sun, the Moon, the five planets, and a crystal sphere of stars. The heavens are perfect and unchanging and the irregular behavior of the comets meant that they had no place in the heavens according to Aristotle.

If comets are nearby objects in the atmosphere, then the observations of comet movements from two different locations on Earth will yield different perspectives of the object, with their trajectories projecting to different parts of the sky. In November 1577, a long-tailed comet traveled through the skies of Europe. For 2 months, Tycho Brahe (1546–1601), a Danish astronomer who made careful and accurate observations of celestial objects, followed the movement of the comet nightly. By comparing the positions of the comet against the stellar background observed from his observatory in Hveen (near Copenhagen) to those observed from Prague, Tycho Brahe was able to conclude that there was no observable difference and therefore the comet had to be far away, at least beyond the orbit of the Moon.

S. Kwok, *Stardust*, Astronomers' Universe, DOI 10.1007/978-3-642-32802-2_18,
© Springer-Verlag Berlin Heidelberg 2013

Fig. 18.1 Ancient perception of the evil nature of comets.
The ancient image of comets is illustrated in this woodcut showing a fourth century comet from Stanilaus Lubienietski's Theatrum Cometicum (Amsterdam, 1668). The comet was perceived to be responsible for destructions on Earth. Photo credit: NASA/JPL

This implies that the comet was a celestial object that appeared from nowhere and had in fact crashed through the crystal sphere of the fixed stars of Aristotle.

It was not until the seventeenth century through improved determination of comets' trajectories that they were conclusively found to be objects revolving round the Sun, much like the planets. Isaac Newton showed with his theory of gravity that objects under the gravitational influence of the Sun can travel in elliptical (as the planets do), parabolic or hyperbolic orbits. Comets can easily belong to the class of objects having the latter two categories of motion. The triumphant confirmation of the prediction by Edmond Halley (1656–1742) of the return of the comet now named after him in 1758 showed without doubt that at least some comets follow highly elliptical orbits around the Sun. Halley based his prediction on the observation that the comets of 1531, 1607, and 1682 all had similar orbits and therefore could be the same object returning every 75–76 years. Other comets of non-periodic nature were found to have parabolic or hyperbolic orbits determined by the gravitational attraction of the Sun (Fig. 18.2).

For most of the twentieth century, comets were often discovered by amateur astronomers who kept a keen watch on the sky. With a good knowledge of the star field, the sudden appearance of a new object may signal the arrival of a new comet. With modern robotic telescopes that are designed to constantly survey the sky, comets can be recognized by computer programs and over 100 comets are now routinely discovered every year. An early example of such a sky survey by a robotic telescope is the Lincoln Near Earth Asteroid Research (LINEAR) project conducted at the Massachusetts Institute of Technology.

In spite of the large number of comet discoveries every year, there still remains the question of where comets come from. Are they interstellar objects captured by the Sun, or are they part of our primordial Solar System? Proponents of the former

Fig. 18.2   Comet Hale-Bopp.
Comet Hale-Bopp of 1997 is
one of the most spectacular
comets in recent years. It was
visible to the naked eye for
over 19 months. The Comet
Hale-Bopp will return in
2,400 years. This picture was
taken on 1997, March 8 at the
Lowell Observatory,
Flagstaff, Arizona. Photo
credit: Michael F. A'Hearn

theory argued that since comets have very eccentric orbits and are inclined at random angles from the orbital plane of the planets (the ecliptic), they cannot belong to the same family. The origin of comets remained a mystery until the middle of the twentieth century when Jan Hendrik Oort (1900–1992) proposed that a large spherical cloud was the reservoir of comets. This cloud extends from 3,000 to 100,000 astronomical units (A.U., the distance between the Earth and the Sun) from the Sun and is now commonly referred to as the Oort Cloud. The outer edge of the Oort Cloud extends one third of the way to our nearest stellar neighbor, Proxima Centauri. A population of comets is held in this reservoir until a member's orbit is perturbed by the passage of nearby stars and it ventures into the inner Solar System. While on this journey, its proximity to the Sun causes the comet's icy surface to evaporate. For the other dormant comets that reside in the Oort Cloud, the Sun is too far away and its radiation too weak to cause any harm. Current estimates put the number of comets in the Oort Cloud at about 100 billion, but it could be as many as tens of trillions.

More detailed studies of the orbits of comets suggest that the Oort Cloud is not the only source of comets. In the inner part of the spherical Oort Cloud is a cloud in the form of a flattened disk. This disk extends to 50 A.U. and is named the Kuiper Belt in honor of Gerard Peter Kuiper (1905–1973), who together with Kenneth Essex Edgeworth (1880–1972), predicted its existence. The Kuiper Belt is probably the remnants left over from the formation of the Solar System, where the inner part condensed into the major planets, leaving the thousands of small, icy objects stranded in the outer parts. In spite of their large volume, the total amounts of mass in the Kuiper Belt and the Oort Cloud are small. The Kuiper Belt and the Oort Cloud are believed to contain respectively 0.1–1 Earth mass and 1–250 Earth masses in bodies of sizes 0.5–500 km.

The first Kuiper Belt Object (KBO) was discovered in 1992, and now we believe there are over 100,000 KBOs with sizes larger than 100 km populating the Kuiper Belt. In 1992, David Jewitt of the University of Hawaii (now at UCLA) and Jane Luu (now at Lincoln Laboratory), discovered the first KBO with the University of Hawaii 2.2m telescope on Mauna Kea (Fig. 18.3). Jane Luu was a Vietnamese refugee who came to the U.S. in 1975 at the age of 12. Her family moved from North Vietnam to South Vietnam in 1955 and her father worked as an interpreter for the U.S. Army. When the South Vietnamese government was collapsing in 1975, the Luu family escaped on a cargo plane during the last days of the war. Her family stayed in the refugee camp for one month and then joined her aunt in Kentucky. In 1976, Jane Luu moved to Ventura, California, to join her father, who had been working as a bookkeeper. Upon graduation from high school in Ventura, she attended Stanford University as an undergraduate and later did graduate studies at the Department of Earth, Atmospheric, and Planetary Science of the Massachusetts Institute of Technology. It was at MIT that Luu began to work with David Jewitt who was an assistant professor at the time.

For their discovery of the first object in the outer Solar System, Jewitt and Luu were awarded the Shaw Prize in astronomy in 2012. The Shaw Prizes were created by the Shaw Foundation established by Sir Run Run Shaw, a media mogul in Hong Kong. The Shaw Foundation gives out prizes annually in the areas of astronomy, life science and medicine, and mathematical sciences, each carrying a value of 1 million USD. In the same year, Jewitt and Luu were also recognized together with Michael E. Brown of the California Institute of Technology with the Kavli Prize for astrophysics. The Kavli Prizes are given out by a partnership of Norwegian Academy of Science and Letters, the Kavli Foundation, and the Norwegian Ministry of Education and Research and are awarded biennially in the areas of astrophysics, nanoscience, and neuroscience.

The KBO Jewitt and Luu discovered, officially designated by the unromantic name of 1992 QB1, was found to be similar to Pluto. Jewitt and Luu, however, nicknamed their object "Smiley" after the British spymaster Geoge Smiley in John Le Carre's spy novels. The discovery of further KBOs showed clearly that Pluto belongs to this new class of Solar System objects and eventually led to the downfall of Pluto's status as a planet. The early 1990s was not a time when the study of the Solar System was considered "sexy", and Jewitt had a hard time convincing the telescope time allocation committee to award precious telescope time to him to search the outer Solar System on the premise that it might not be empty. Eventually he had to resort to piggy-backing on the telescope time that he was awarded to use for other projects. After struggling for 5 years, their effort handsomely paid off when Jewitt and Luu found a moving object located at a distance of more than 40 A.U. from the Sun. This was the beginning of an avalanche of discoveries leading to over 1,000 KBOs known to date.

The visual brightness of KBOs is due to reflected sunlight. The farther away they are, the less sunlight they can intercept. The detection of KBOs therefore relies on the interception of their reflected light. So the brightness of a KBO fades very rapidly with distance. Beyond a certain distance, KBOs become very difficult to find, and their detection requires larger and larger telescopes. The way they are discovered is by taking successive pictures over a certain period. If an object is seen to have moved relative to the stars, it is probably a Solar System object. From its trajectory, one can determine its orbit in the Solar System. The size of the object can be determined from its visual brightness. The combination of these pieces of knowledge can tell us whether an object is a KBO.

Fig. 18.3 Discovery image
of the first KBO.
The object pointed to by the
arrows is the first KBO, 1992
QB1. The four frames show
that the object is moving
relative to the fixed stars.
Image credit: David Jewitt

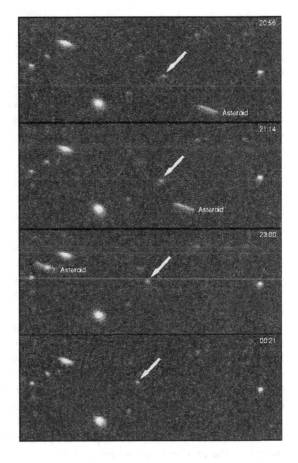

This kind of search requires a constant survey of the entire sky. We need telescopes with a wide field of view so that pictures can be taken covering as large a fraction of the sky as possible. The same parts of the sky are visited time and time again in order to obtain the temporal information, seeking out objects that are moving against the star field. At present, there are two major telescope projects designed with this kind of survey in mind. One is *the Panoramic Survey Telescope and Rapid Response System (Pan-STARRS)* project. *Pan-STARRS* consists of a system of four 1.8-m telescopes located on the Mauna Kea and Haleakala mountains in Hawaii. From the latitude of Hawaii (~20° N), *Pan-STARRS* can cover approximately three quarters of the entire sky. The project is supported by the U.S. Air Force with the goal of identifying near-Earth objects, small asteroids that cross the Earth's orbit and may have a chance of hitting the Earth. In addition to near Earth objects, the telescope will discover many new asteroids, satellites, comets, and KBOs.

The second project is the *Large Synoptic Survey Telescope (LSST)* of the *National Optical Astronomy Observatory* of the U.S.A. *LSST* is an 8.4-m telescope to be located on 2,682-m Cerro Pachon in Chile. Its main instrument is a 3.2 gigapixel digital camera, capable of taking 15-s exposure pictures every 20 s. It is expected to take over 200,000 pictures each year, therefore providing a rapid survey of the entire sky.

When these two telescopes go into full operation, our knowledge of the Solar System, both inner and outer, will be greatly expanded. It is amazing that while astronomers have been trying to reach out to ever distant galaxies billions of light years away, we still do not have a good understanding of our own backyard—the Solar System.

So now we know comets come from a deep freeze store in the outer regions of the Solar System, but what are they made up of? The observations of the cometary tail suggest that comets are gaseous in composition. The increasing brightness of comets as they approach the Sun points to sublimation as the process of production of the tail. On Earth, we are familiar with sublimation as the way that snow and ice is directly converted to water vapor by sunlight without going through the phase of liquid water. From this experience in our daily lives, astronomers speculated that comets must have an icy core from which the gaseous materials are released during the comet passage. This basic idea forms the popular model of comets being "dirty snowballs" by Fred Whipple (1906–2004).

Because this solid core (called the nucleus) is shrouded inside the gaseous head (called coma) and tail of the comet, its physical size is difficult to ascertain. However, from the absence of any discernible gravitational effect on the orbits of planets (including the Earth) in close encounter with comets, astronomers were able to show that comets have masses much smaller than those of planets. The opportunity to directly measure the size of comets occurred on May 8, 1910, when Halley's comet's path crossed the face of the Sun. The fact that no dark spot could be seen on the solar disc puts an upper limit of 100 km for the size of the nucleus. The precise measurements of the sizes of cometary nuclei were made possible by space flights. The return of Halley's Comet in 1986 was met by a fleet of spacecrafts which found its nucleus to be of irregular shape and to have a size of $15 \times 7.5 \times 7.5$ km. In 2001, NASA's *Deep Space 1* spacecraft approached within 2,171 km of Comet Borrelly and found its nucleus to be elongated measuring 8×3.2 km. The most famous recent examples are the pictures taken of the Comet Wild 2 by the spacecraft *Stardust* as it flew by the comet on January 2, 2004, at the close approach of 236 km. The rugged surface of the nucleus of Wild 2 measures about 5 km across (Fig. 18.4). These observations show that comets are intrinsically very much like asteroids (Fig. 2.2), with the exception that the former have outgassing activities whereas the latter are dormant.

The prize for the most dramatic action probably should go to the *Deep Impact* spacecraft, which sent an impactor to collide with Comet Tempel 1 on July 4, 2005. From the picture taken by the impactor before collision, the comet nucleus was found to have a size of 6 km. After the collision, the plumes ejected from the impact site were extensively observed by ground-based telescopes, and a variety of molecules and dust particles were identified (Fig. 18.5).

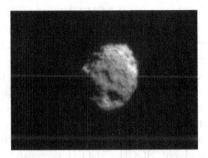

Fig. 18.4 Image of Comet Wild 2.
The image of the comet Wild 2 taken by the *Stardust* spacecraft on January 2, 2004. A number of features resembling valleys, hills, and craters can be seen on the surface of its 4 km nucleus. Unlike surface features on asteroids and the Moon, the basins are not due to external impacts

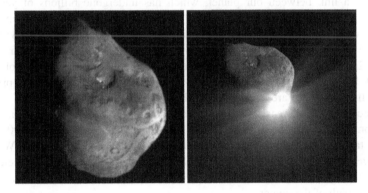

Fig. 18.5 Comet Tempel 1.
Left: The image of the comet Tempel 1 taken approximately 5 min before Deep Impact's probe smashed into its surface. The nucleus of this comet was illuminated by the Sun, which was shining from the right. The circular features on the surface are impact craters. The probe was about to crash into the region between the two carters near the lower center of the comet. Right: Image of comet Tempel 1 taken 67 s after impact. Image credit NASA/JPL-Caltech/UMD

Although the general visual light from comets is due to reflected sunlight, detailed observations can reveal the chemical composition of the object. Astronomical spectroscopic observations have been possible since the late nineteenth century. Since molecules emit light at specific wavelengths, exact molecular species can be identified by comparing the cometary spectrum with laboratory spectra of known molecules. Optical observations in the early twentieth century have led to the discovery of molecules such as C_2 and CN. However, many molecular species radiate only in the infrared or submillimeter wave parts of the spectrum and their discoveries had to wait until the development of the observational techniques in the 1970s. In addition to water vapor (H_2O), we now know that comets contain a large variety of molecules, including carbon monoxide (CO), carbon dioxide (CO_2), formaldehyde (H_2CO), ammonia (NH_3), hydrogen cyanide (HCN), methanol

(CH_3OH), acetylene (C_2H_2), acetonitrile (CH_3CN), isocyanic acid (HNCO), hydrogen sulfide (H_2S), etc. The list of detected gas-phase molecules goes on to include some quite complex organics such as acetaldehyde (CH_3CHO), methyl formate ($HCOOCH_3$), and formamide (NH_2CHO).

These gaseous molecules are detected with astronomical telescopes as they are released from comets as the result of evaporation by sunlight. Modern infrared spectroscopic techniques can directly detect the ices on the comet surface as well. The most common form of ice is water ice, but ices of the natural gas methane (CH_4), ethane (C_2H_6), alcohols (CH_3OH), acids (HCOOH), formaldehyde (H_2CO) and antifreeze (ethylene glycol, $HOCH_2CH_2OH$) have also been detected. Some of these ices, e.g., CO and CH_4, are extremely volatile and will evaporate under conditions of a few degrees above absolute zero. Michael Mumma of the Goddard Space Flight Center argues that these comets are entirely pristine and their chemical composition has remained unchanged for 4.5 billion years. Comets therefore represent a link between our planet, which has undergone billions of years of physical and chemical processing, and the world outside the Solar System. The study of comets will tell us what the original condition of the Solar System was like.

Since these molecular species have all been detected in interstellar space (Chap. 17), it is natural for us to speculate that comets inherited their molecular content from outside the Solar System. Michael Mumma believes that pristine interstellar molecules are trapped inside the ices in the nuclei of comets.

In recent years, increasing attention has been paid to the non-gaseous or solid-state components of comets. Figure 18.6 shows the Comet McNaught of 2007. The curved tail of the comet is due to dust particles illuminated by the Sun. When a comet gets close to the Sun, heat from the Sun causes gas and dust on the nucleus to be released. Pushed by light pressure from the Sun, the dust particles form a tail along the path of the comet.

In Chap. 6, we have already mentioned that dust was discovered in comets in the 1960s through infrared observations. These dust particles are micrometer-size particles, with silicates being the first constituent identified. In addition to the amorphous silicates, detected through their characteristic emissions at 10 and 18 μm, observations from the *Infrared Space Observatory* have also identified silicates in a crystalline form. So comets are more than just ice balls, but in fact contain various minerals. In the 1986 missions to Halley's Comet, the two Soviet *Vega* spacecrafts and the European Space Agency's *Giotto* spacecraft found that the dust particles released by the comet include magnesium, silicon, calcium and iron, the common elements found in rocks, as well as carbonaceous compounds made up of carbon, hydrogen, oxygen, and nitrogen, the basic ingredients of organic substances.

Most excitingly, the *Stardust* spacecraft flew to Comet Wild 2, collected almost one million cometary dust particles and brought them back to Earth on January 15, 2006 after a seven-year journey. The capsule containing the sample landed in the Great Salt Lake desert in Utah and was flown to NASA Johnson Space Flight Center in Houston, Texas, for safe keeping. Laboratory analysis of the returned samples found an abundance of amorphous and crystalline silicates, as well as a wide range

Fig. 18.6 Comet McNaught.
A picture of Comet McNaught as observed on January 18, 2007 during the twilight hours in Cerro Pachon, Chile. The tail of the comet can be seen extending over 30°. Photo credit: Marie-Claire Hainaut, Gemini Observatory

of organic compounds. The mix of organics in this sample is not too different from the insoluble organic matter found in meteorites (see Chap. 15). This raises the question of whether the organic matter found in comets and meteorites share a common origin. Even more interesting is whether they inherit these organic substances from the interstellar medium?

Although KBOs are faint and difficult to study by spectroscopy, astronomers have used modern large telescopes such as the 10-m *Keck*, 8-m *Gemini*, 8-m *Subaru*, and the 10-m *Very Large Telescope* to study a few dozen KBOs. Infrared signatures of methane ice and ammonia ice have been found. There is also evidence that complex organics may be present on their surfaces based on the colors of the objects. All these detections point to the fact that these primitive objects already possessed organics very early in the Solar System's history.

Even after a comet is long gone, the debris left behind by the comets can still be felt on Earth. When the orbital motion of the Earth passes through the remnants of a comet, the dust particles rain down on Earth in a spectacular phenomenon known as a meteor shower. From an observer's point of view, a large number of meteors emerge on the sky with an apparent common origin - for example, the Perseid meteor shower that occurs with maximum intensity annually on August 12–13. From his observations of the Perseid meteors in 1864 and 1866, the Italian astronomer Giovanni Schiaparelli (1835–1910) concluded that the Persied meteors are associated with the remnants of Comet 109P/Swift-Tuttle. Similarly, the Leonid meteor shower that occurs on November 17 is tied to Comet 55P/Tempel-Tuttle. The Quadrantids observed in late December and early January are named after the meteor showers radiating from the now defunct constellation Quadrans Muralis. In the early twentieth century when the International Astronomical Union reorganized the constellations, the constellation Quadrans Muralis was eliminated and incorporated within the constellations Boötes and Draco. The Quadrantids showers are traced to a minor planet named 2003EH$_1$, which was probably part of a comet that broke up about 500 years ago. It has been suggested this comet was actually observed by Chinese astronomers in the 1490s. After 500 years, the remnants of the comet are still around, and every year the Earth captures some of them and they make spectacular entries into the atmosphere in the form of meteors.

Meteors are also commonly called "falling stars" as they are bright objects that seem to move from the sky to Earth (Chap. 2). In ancient Chinese folklore, meteors were foretellers of the imminent demise of significant people such as emperors and kings, who were believed to be associated with the stars. We now know that these showers are the result of the Earth encountering the graveyard of comets. So the death analogy is not unreasonable. The main difference is that a comet can be broken up into many pieces with each remnant piece seen as a meteor. Although they can be as bright as stars, they are much, much smaller than stars and their brightness is due to their proximity to us, being in the Earth's atmosphere instead of light years away.

As soon as astronomers realized that comets are more than just ice balls composed of mainly inorganic substances, they began to raise questions about the possible implications of comets delivering ingredients for life to Earth and other rocky planets. Armand Delsemme of the University of Toledo in Ohio was one of the most vocal advocates. The delivery of extraterrestrial organic materials to Earth can take one of several forms: (1) they can come in a dramatic form as part of a comet impact; (2) as part of asteroid impact with remnants in the form of meteorites; or (3) they can rain down on Earth as part of the daily flux of interplanetary dust particles. One important question is whether this organic matter can survive intact without being altered significantly by the impact. Since the impactor must enter the Earth at a minimum velocity of 11.2 km/s (the escape velocity for the Earth, see Appendix F), the resultant impacts will generate high temperatures. If the kinetic energy of a comet is completely converted to heat upon impact, the temperature can reach tens of thousands of degrees, implying a complete breakdown of organic substances. Others have argued that the period of exposure to high temperature is brief, and the organics could well be preserved. Experiments performed in the laboratory simulating impact conditions have yielded conflicting results, with some even finding that the impacts increase the complexity of biological molecules. Certainly more experimental studies are necessary.

Proponents of the impact destruction argument assume that these external impactors are hitting the solid Earth. However, the Earth's surface is made up of mostly water. The survival of the external bodies would be much easier if the impact occurred on water. In fact, some scientists believe that all the water in our oceans was delivered to us by comets (Chap. 19). As we have read in this chapter, comets are made up of chunks of ice with organic substances mixed in. In the process of bringing water to Earth, the organic components could have easily been carried with them.

Comets have a spectacular appearance and their mysterious arrivals and departures have long fascinated ancient men. We now know that they originate from the outer Solar System and we become aware of them when they venture into the inner Solar System. They have a solid core with an ice mantle, which evaporates when they get too close to the Sun. The dust and gas released create the magnificent tails that we see. Beneath the heavenly beauty of the comet lies a rocky nucleus, which is very similar to the Earth's surface. Common minerals and simple gaseous molecules make up the ejecta from the comet. In addition, comets also carry complex organic compounds, which they probably inherited from stellar or interstellar sources. Remnants of comets are captured by the moving Earth and their entries into the atmosphere create the optical phenomenon of meteor showers. Occasionally a comet hits the Earth, and such impacts have tremendous effects on Earth. Some of these impacts are detrimental to life on Earth and some may be beneficial. The latter is the topic of the next chapter.

A brief summary of this chapter
Comets come from the outer Solar System and carry complex organic compounds, possibly from stars or interstellar sources.

Key words and concepts in this chapter
- Origin of comets
- Oort Cloud as a reservoir of comets
- Search for Kuiper Belt objects
- Pluto lost its status as a planet
- Size and composition of cometary nuclei
- The Deep Impact mission
- Molecules and solids in cometary tails
- Comets as reservoir of interstellar molecules
- Origin of meteor showers
- Return of samples from Comet Wild 2 by the *Stardust* mission
- Organic content in comets
- Comets as delivery agents

Questions to think about
1. The ancient people saw comets as a transient phenomenon which is different from the peaceful and everlasting nature of celestial bodies. Many cultures saw comets as bad omens. Based on our discussions on the relationship between comets and Earth, what is your opinion on whether comets are good or bad for the human race?
2. The visual brightness of comets is derived from reflected sunlight whereas its infrared brightness is due to self radiation. Which do you think is a better technique to observe and understand the nature of comets? What are the advantages and disadvantages of the two techniques?

(continued)

3. Now with the existence of the Kuiper Belt and the Oort Cloud known, the Solar System seems to be much larger than when it was defined as the boundary of the planetary orbits. How can we define precisely the boundary of the Solar System?

Chapter 19
Where Do Oceans Come from?

What makes the Earth a livable planet? One simple, but commonly accepted answer is "a planet that allows liquid water to be stably present". This statement puts a lot of weight on water being the essential element for life. Some scientists have argued whether it is necessary to require surface water, or whether subsurface liquid water is enough. Nevertheless, few dispute that so long as life is carbon-based, it will require liquid water to begin, to survive, and to flourish. Given the importance of water, is it a coincidence that the Earth is blessed with so much liquid water?

About three quarters of the Earth's surface is covered by water in the form of oceans, seas, rivers, and lakes. The amount of liquid form of water far outweighs the amount of water in solid form as (ice and snow) and in the gaseous phase (as water vapor in the atmosphere). The average depth of the ocean is 3.7 km, with the deepest at 10.9 km at the Marianas Trench in the northwest Pacific. To put this in perspective, this depth is more than the height of Mount Everest of 8.8 km. It is estimated that ocean water takes up a total volume of 1.37 billion cubic kilometers and has a total mass of 1.4×10^{21} kg. The atmosphere pales in comparison: if the entire atmosphere condensed into ice, it would only be 10 m thick.

The oceans are also a mysterious place. Even in the clearest waters, sunlight cannot penetrate more than 10 m of water and at 100 m less than 1 % of the sunlight remains. This means that most of the ocean is in permanent darkness. Plants and organisms that rely on sunlight for photosynthesis can only survive near the surface. Unlike the continents, which have been extensively explored by men, we have been totally ignorant of the ocean floors until recently. In the several thousand years of human history, different races of people have migrated over the continents of the Earth. Early in human history, people moved out of Africa to Europe and Asia, and later Asiatic people crossed over the Bering land bridge and populated the continents of the Americas. Over sea, the Polynesians have traveled across the Pacific Ocean and in recent times Europeans have crossed the Atlantic to reach the Americas. In the last 200 years, we have extensively explored the surface of the Earth and created accurate maps of the world. However, the existence of submerged mountains, plateaus, basins and deep trenches only became known after the exploration journey of *H.M.S. Challenger* in 1872. The *Challenger* (Fig. 19.1),

S. Kwok, *Stardust*, Astronomers' Universe, DOI 10.1007/978-3-642-32802-2_19,
© Springer-Verlag Berlin Heidelberg 2013

Fig. 19.1 H.M.S.
Challenger.
A painting of the H.M.S.
Challenger by William
Frederick Mitchell

launched on February 13 1858 by the British Navy, conducted a number of military missions. Commissioned by the Royal Society of London for this scientific mission, all but two of the guns were removed from the ship in order to make room for scientific equipment. Commanded by Captain George Nares and scientifically supervised by Charles Wyville Thomson of the University of Edinburgh, the *Challenger* spent 713 days at sea and traveled almost 130,000 km surveying and studying the oceans.[1] The scientific results gathered from this journey provided the foundation for the modern discipline of oceanography, the study of the Earth's oceans. These expensive expeditions in the nineteenth century are equivalent to the space exploration missions that we have today. Modern techniques to map the seafloor topography include employing sound waves from ships as well as radar from satellites.

Oceans also have been around for a long time. Carbonate sedimentary rocks, which are formed in water, as old as 1 billion years have been found. The discovery of fossils of bacteria and algae dating back to 3.3 billion years ago can also be used as evidence of existence of water at that time. However, beyond the indirect evidence above, there is no record of the early presence of water on Earth.

Water is so prevalent on Earth that we take it for granted. How many of us have thought of the question: where did the oceans come from? Among the early theories is the volcano origin theory by the Swedish geologist Arivd Gustaf Högbom (1857–1940), who in 1894 proposed that oceans were condensed from volcanic gases containing steam and carbon dioxide. At first, the answer seemed obvious: water on Earth must have come together with the rocky materials when the Earth formed. As the Earth heated, the rocks melted and the oxygen and hydrogen atoms

---

[1] I am sure it did not escape the notice of the readers that the space shuttle *Challenger* was named after this exploration ship. Similarly, the space shuttle *Endeavour* was named after the British ship *HMS Endeavour*, which Captain James Cook took on his first voyage (1678–1771). The shuttle *Discovery* was named after *H.M.S. Discovery*, one of the ships on Cook's third voyage (1776–1780).

in minerals were released through volcanic activities. The released gas then condensed into oceans.

However, we now know that the Earth suffered from a giant impact with an object the size of Mars, which ripped out a major part of the Earth's outer parts from which the Moon was formed (Chap. 14). Whatever oceans that might have existed on Earth before the impact must have been evaporated as a result. In spite of its obvious importance, the question "how did Earth get its water" has not been satisfactorily answered as of the early twenty-first century. Scientists have switched from the traditional view of water coming out of the Earth's interior through volcanism, to various models of water being deposited on the Earth's surface. The view that terrestrial water was delivered by comets was promoted by Christopher Chyba of Cornell University and Armand Delsemme of the University of Toledo. Comets have the advantage that they contain a lot of ice. If there were enough comets hitting the Earth during its early history, the amount of water on Earth could be easily accounted for.

From the craters on the Moon, one can also reconstruct the rate of bombardments, which only diminishes (by about a factor of 1,000) after one billion years. The primary agents of bombardment are comets. From the crater record on the Moon, one can estimate that the Moon was hit by comets with sizes ranging from 5 to 500 km. The same bombardments must have also struck the Earth. The larger gravity of the Earth compared to the Moon in fact means that the rate must be larger for the Earth than for the Moon. The gravity has an even more important effect: the Earth was able to retain most of the captured materials whereas the weaker gravity of the Moon could not. From the composition of the comets, we can tell that this period of giant bombardment must have brought to Earth more than 10 times the amount of water in the present oceans, and more than 1,000 times the amount of gas in our present atmosphere. As it cools, the cometary water vapor condensed to form the oceans. The solar ultraviolet light dissociated the cometary ammonia (NH_3) into nitrogen (N_2) and hydrogen (H_2). Since the Earth's gravity is too weak to keep the hydrogen, only the nitrogen remains in our present atmosphere.

Comets are not the only source of water in the Solar System. In 2006, Henry Hsieh and David Jewitt of the University of Hawaii discovered a new class of comets. They are called the Main Belt Comets because they are located in the asteroid belt but have outgassing activities similar to comets. They are believed to have reservoirs of ice hidden under their surfaces and when they get close to the Sun, the ice begins to sublime and water vapor is released. As they have appearances like those of asteroids, they are hard to pinpoint. It is estimated that there are thousands of Main Belt Comets lying within the asteroid belt.

To further compound the issue, the amount of water that we can observe and measure in the oceans may not represent the total amount of water on Earth. It may not even represent the majority. There is water embedded in the rocks, both in the Earth's crust and mantle. A detailed breakdown of water content in the Earth is shown in Table 19.1. As we can see, our knowledge is highly uncertain. The best estimates of the amount of water in the mantle range from one-tenth to 1.5 times the

Table 19.1 Total water content on Earth

Sources	Mass of water (10^{18} kg)
Oceans	1371.3
Marine sediment porewater	180
Marine basement formation water	26
Ice	27.8
Continental groundwater	15.3
Lakes, rivers, soils	0.192
Total hydrosphere	**1,621**
Shales	221
Continental carbonates	2.56
Evaporites	0.42
Marine clays	7.56
Marine carbonates	0.504
Total sedimentary rocks	**232**
Organic matter	1.36
Continental metamorphic rocks	36
Oceanic (igneous) crust	41
Total hydrosphere and crust	**1,931**
Mantle	
Serpentinite in oceanic lithosphere	11–27
Upper mantle	
To 410 km	74
410–670 km	56–930
Lower mantle	59–1,470
Total mantle	200–2,500
Total Earth (except core)	
Low estimate	2,130
Mid estimate	2,210
High estimate	4,430

Content of table derived from Mottl, M., Glazer, B., Kaiser, R., & Meech, K. 2007, *Chemie der Erde/Geochemistry*, **67**, 253

amount of water in the oceans. In the case of water in the core, it is even worse. It is estimated that the amount of water in the core could be as high as 100 times the amount in the oceans.

Biological activities may also have contributed toward the formation of oceans. James Lovelock, in his Gaia hypothesis, postulates that the living and non-living parts of the Earth form an integrated, interacting system. The name of this hypothesis is based on the mythical Greek goddess who is Mother Earth. In his model, the Earth's crust, the oceans, the atmosphere, and the biosphere are all part of a single entity. Changes in one component will lead to feedback in the other components. Within the Gaia framework, both our present atmosphere and oceans were created as a result of life. The early Earth acquired its water from the Earth's crust. After the emergence of life from this small pond, oxygen released from bacteria substantially increased the amount of oxygen in the atmosphere and these oxygen atoms combined with hydrogen to form water. While the evolution of the Earth's atmosphere as a result of life is widely accepted, the origin of the oceans is still under debate.

As we read from the last chapter, comets contain not just ice and inorganic dust, but also organic compounds. During the period of giant bombardment in the first one billion years, comets must have brought in excess of 10^{18} tons of prebiotic molecules to Earth. The possible influence of these prebiotic molecules on the development of life on Earth will be explored in the following chapters.

A brief summary of this chapter
The possibility that water on Earth was carried here by comets.

Key words and concepts in this chapter
- The role of water in making a planet habitable
- Origin of water on Earth
- Origin of the Moon
- Total amount of water on Earth
- Delivery of water to Earth by comets
- The Gaia hypothesis and an integrated model of the Earth

Questions to think about
1. Before you read about the comet origin of ocean water, where did you think the water on Earth came from?
2. The earth's crust is full of rocks. Where did these rocks come from?
3. Why is it so difficult to determine how much water we have on Earth?
4. Mountains rose, oceans formed, glaciers came and went, and continents drifted over the history of the Earth. What is the evidence that these events happened?
5. Earth has a long history. How do scientists know what events happened at what time? How can we be so precise in dating ancient events that happened on time scales of millions and billions of years ago?
6. The chemical composition of our current atmosphere is the result of life. However, many living organisms now depend on the atmosphere to survive. How is this balance maintained?
7. What do you think of the Gaia hypothesis? Are the physical and biological components really totally integrated as put forth by this hypothesis?

Chapter 20
Playing God with Primordial Soup

Since the original Miller–Urey experiment where a mixture of methane, ammonia, hydrogen, and water under spark discharges was found to be able to produce amino acids (Chap. 1), there have been many follow-up experiments to test the synthesis of organic matter in simulated primitive Earth atmospheres and different energy inputs such as ultraviolet light or heat. A sample used by Stanley Miller in 1958 was retrieved from archive and re-analyzed with modern instruments. This sample contains a mixture of methane, ammonia, H_2S and carbon dioxide. The inclusion of the element sulfur was intended to simulate the condition under volcanic plumes. Putting this mixture under a spark discharge, a group at the Scripps Institution of Oceanography in San Diego, California, found a total of 23 amino acids as products. This new experiment confirmed and extended the conclusion of Miller that prebiotic molecules can be produced on Earth under suitable conditions.

With the discovery of gas-phase molecules in interstellar space, there have been a number of attempts to study the possibility of organic synthesis on the surface of solid-state dust grains under interstellar conditions. Starting with simple ices such as water (H_2O), carbon monoxide (CO), methane (CH_4), ammonia (NH_3), methanol (CH_3OH), and acetylene (C_2H_2) ices at low (~10 K) temperatures, an organic substance called "yellow stuff" can be produced by subjecting the sample to ultraviolet irradiation and warming it afterwards. When this substance was carried into space and subjected to 4 months of solar radiation, its color changed from yellow to brown. Mayo Greenberg (Fig. 20.1) (1922–2001) of Leiden University in the Netherlands called this substance organic refractory matter, based on its infrared spectrum showing the 3.4-μm feature due to the methyl ($-CH_3$) and methylene ($-CH_2$) aliphatic groups, as well as features suggestive of alcohols and carboxylic acids.

S. Kwok, *Stardust*, Astronomers' Universe, DOI 10.1007/978-3-642-32802-2_20, © Springer-Verlag Berlin Heidelberg 2013

Fig. 20.1 Mayo Greenberg.
Jerome Mayo Greenberg was
a pioneer in astrochemistry.
His laboratory simulations
show that complex organics
can be synthesized on the
surface of dust particles under
space conditions

Greenberg was a native of Baltimore, Maryland. He entered Johns Hopkins University in Baltimore at the age of 15, and started graduate school at the age of 17. His studies were interrupted by World War II when he worked at Langley Field, Virginia, on the problem of airflow over aircraft wings. After graduation in 1948, he taught at the Rensselaer Polythechnic Institute in Troy, New York. His interest in interstellar dust began with his work with Henk van de Hulst at Leiden University in Holland. In 1975 at the age of 53, Greenberg accepted a position as Chair of Laboratory Astrophysics at Leiden University. In Leiden, Greenberg established a laboratory to explore the role of ice mantles on solid grains in fostering chemical reactions. His most famous experiment on organic refractory matter led Greenberg to champion the idea that complex organic molecules and even biogenic species can be produced on grain surfaces in space. These ideas were considered radical and probably would have had difficulty receiving support under the normal peer review granting system at that time. It was fortunate that in Leiden he had the independent funding necessary to pursue these ideas.

After the pioneering work by Greenberg, several groups performed simulations of chemical reactions on ice surfaces, hoping to synthesize complex organic molecules under conditions similar to those in interstellar space. These groups include those led by Louis d'Hendecourt (a former student of Greenberg) of Centre National de la Recherche Scientifique (CNRS) in France, and Louis J. Allamandolla in NASA Ames Research Center in California. Max Bernstein, a chemist from Brooklyn, New York, began his work on interstellar chemistry with Louis Allamandola and Scott Sandford at NASA Ames. They bombarded ices containing water (H_2O), methanol (CH_3OH), carbon monoxide (CO), and ammonia (NH_3) (mixing them in ratios consistent with ices in molecular clouds) with ultra-violet light. In a paper with Allamandola and Sandford in 1995, Bernstein reported that his experiment created a rich variety of moderately complex organic molecules such as ethanol (CH_3CH_2OH), formamide ($HC(=O)NH_2$), acetamide ($CH_3C(=O)NH_2$), nitriles ($R-C\equiv N$), and hexamethylenetetramine (HMT, $C_6H_{12}N_4$), as well as more complex unidentified organic materials that include amides ($H_2NC(=O)-R$), ketones (R–C(=O)=R'), and polyoxymethylenes (POM, $(-CH_2O-)_n$), where R represents a generalized alkyl group.

HMT, which contains four nitrogen atoms per molecule, was one of the major organic residues produced by the experiment. It has been known since the 1970s that HMT forms spontaneously at room temperature when pure ammonia and formaldehyde are mixed. Since it is known that mixing HMT with concentrated acid can yield amino acids, the formation of this molecule in the interstellar medium has great significance.

This line of experiments using ultraviolet light to transform amorphous water ice was continued by the Ames and Leiden groups. In two papers published back to back in *Nature* magazine in 2002, these two groups reported that by warming, ice sublimation leaves behind an organic residue which contains N-formyl glycine, cycloserine, and glycerol. After hydrolysis (that is by adding water and mixing with the residue), glycine, alanine, serine, glycerol, ethanolamine and glyceric acid are observed. Similar experiments using different initial ingredients also yielded many amino acids. These results suggested that spontaneous generation of amino acids in space is possible.

Similar efforts were also made by a group in Yokohama National University in Japan led by Kensei Kobayashi. They were able to produce nucleic acid bases, which are believed to be essential for the generation of life. Proton irradiation of carbon monoxide, nitrogen and water under simulated interstellar conditions has been shown to be able to yield uracil, one of the four RNA bases (the others being cytosine, adenine, and guanine). Thymine, the base that substitutes for uracil in DNA, has been reported to be present after proton irradiation of a mixture of methane, carbon monoxide, and ammonia.

The significance of these laboratory simulations is that prebiotic molecules such as amino acids need not be produced on Earth. With a suitable medium (e.g., ice covering solid particles ejected by stars) and the injection of energy (e.g., in the form of ultraviolet light, which is in ample supply from nearby stars), complex molecules can be created in space. These experiments therefore have taken a step beyond the Miller–Urey experiment. While the Miller–Urey experiment showed that complex biomolecules can be made from simple ingredients under the conditions of early Earth, the Ames/Leiden experiments show that it is possible under interstellar conditions. The difference in implication is immense. While the Miller–Urey experiment showed that the ingredients of life could be made on Earth naturally, the Ames/Leiden experiment showed that this could be possible ANYWHERE. If life could emerge from these ingredients on Earth, then life could just as easily emerge on another planet elsewhere.

However, even if the Ames/Leiden experiments are accurate reflections of reality, there still remains the problem of taking these grains from the interstellar space to Earth. At the beginning of the twenty-first century, there is an increasing degree of recognition that external delivery may play a strong role in the creation of life on Earth. We know that complex organic compounds are not the exclusive domains of the Earth. In fact, stars can make them with remarkable ease and a huge amount of organics are produced and distributed by stars into the interstellar space

Astrobiology coming of age

Astrobiology (sometimes also referred to as Bioastronomy or Exobiology) is a young science that has received great attention in the international science community in recent years. The International Astronomical Union recognized the importance of this emerging field by creating a new commission (No. 51, Bioastronomy), which has run a conference once every 3 years. A group of chemists and biologists started the International Society for the Study of the Origin of Life (ISSOL) in 1973. ISSOL held meetings on the origin of life once every 3 years but traditionally had little interaction with astronomers. Recognizing recent progress in the field, ISSOL held a joint meeting with IAU Bioastronomy in Montpellier, France in 2011 and will do so again in Nara, Japan, in 2014. In the U.S., NASA began its push in this field when Administrator Daniel Goldin created the NASA Astrobiology Institute in 1995. The bi-annual Astrobiology Science Convention has become the focal point of US activities. In Europe, the European Astrbiology Network Association organizes a workshop every year somewhere in Europe. These astrobiology meetings are now routinely attended by astronomers, chemists, biologists, geologists, and planetary scientists.

every year (Chap. 10). We also have strong isotopic evidence that external stellar materials traveled across the Galaxy and reached the Solar System, and indeed we can hold stardust in our hands (Chap. 16). Analysis of meteorites, interplanetary dust, and comet materials has shown that they are rich in organics and prebiotic molecules. We also know that the Earth, the Moon, and our sister planets suffered from heavy external bombardments during their early history and it is very likely that external organic substances enriched the chemical makeup of the early Earth (Chap. 3). These bombardments are believed to have occurred about 3.8–4.0 billion years ago, and remarkably, life emerged on Earth very shortly after. Was it a coincidence or was the development of life on Earth assisted by externally delivered ingredients?

If indeed the Earth received life-forming materials such as amino acids and nuclear bases from the outside, why does life on Earth only use 20 amino acids in its proteins? Also, why are only four nucleobase pairs used in our DNA and RNA? Laboratory experiments have shown that the other amino acids are perfectly capable of making different proteins, and other nucleobase pairs can function well in enzymes. These questions were asked by Lucy Ziurys of the University of Arizona during the International Astronomical Union symposium on Organic Matter in Space held in Hong Kong in 2008. One possible answer suggested by Ziurys is that life happened so quickly on Earth that it did not have time to try all chemical possibilities. Life just evolved based on whatever chemicals happened to be handily around at the time. There was no design, no careful planning, no detailed experimentation; life just began using the externally supplied ingredients available. Life could have taken another path, resulting in biological organisms that are totally different from what we are familiar with.

A brief summary of this chapter
Laboratory simulations show that the synthesis of biomolecules in space is possible.

Key words and concepts in this chapter
- Simulation of the early Earth in the laboratory
- Synthesis of biomolecules in space
- Survival of biomolecules in space
- External delivery of complex organics to Earth
- Alternative chemical paths to life

Questions to think about
1. Do you think these laboratory experiments are accurate simulation of molecular synthesis in space? How do the actual conditions in space differ from the physical conditions of these simulation experiments?
2. Science fiction always depicts aliens as human-like, e.g., in the TV series "*Star Track*", movies "*ET*" and "*Close Encounters of the 3rd Kind*". Even in non-human-like situations (as in the movie "*Alien*"), they still have a head and several limbs. If there are alternate life forms in space, what do you think they look like?
3. If we were able to make use of the full complement of amino acids to make life, how different would these life forms be compared to those arising from only 20 amino acids?

Chapter 21
Stardust and Origin of Life

In the previous chapters of this book, we have presented evidence that organic compounds are common in the Galaxy. They are in our backyards, as interplanetary dust particles, floating around beyond the Earth in the space between planets. They are in asteroids, the remnants left over from the early Solar System. They are in comets, visitors from the outer parts of the Solar System which carry pristine original materials. They are in planetary satellites, Titan being the prime example where organics are believed to be widespread both in its atmosphere and on its surface. Most importantly, they are part of the bulk components of carbonaceous meteorites, which represent bits and pieces of comets and asteroids that have fallen onto the Earth. It is from the analysis of meteorites that we have obtained the most information on the primordial organics, the creation of which dates back to at least the beginning of the Solar System 4.6 billion years ago.

Organics are also in the surroundings of stars. Simple organics are manufactured by old (AGB) stars and ejected from the stars. In the several thousand years after the end of the AGB phase, complex organics are synthesized by planetary nebulae in large quantities.

The interstellar medium of our Galaxy is filled with clouds of dust and gas. Some of these clouds (called emission nebulae) are associated with new born stars, and shine by atomic radiation after absorbing ultraviolet light from the young stars. The Orion Nebula is the most well-known example of this class. Some are just in the general neighborhood of stars and shine by reflecting star light (called reflection nebulae). The UIR emission features are found in both kinds of interstellar clouds. An example of reflection nebulae showing UIR features is shown in Fig. 21.1.

Beyond the disk of the Galaxy, the UIR features are also found in the diffuse clouds (called cirrus) in the halo of the Galaxy. Since we now know that the UIR features are emitted by aromatic compounds (Chap. 10), this proves that organic matter is everywhere in our Galaxy. Surveys of the Galaxy by balloon experiments and also by the Japanese satellite, *Infrared Telescope in Space* (*IRTS*), have found that the UIR features are present in the general diffuse interstellar medium, not associated with any specific clouds. It is estimated that at least 15 % of the element

S. Kwok, *Stardust*, Astronomers' Universe, DOI 10.1007/978-3-642-32802-2_21,
© Springer-Verlag Berlin Heidelberg 2013

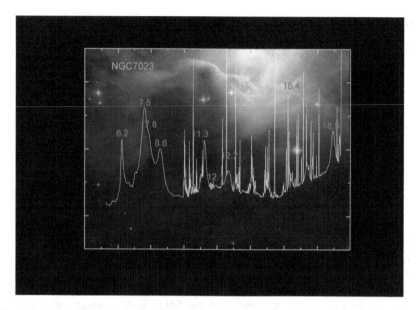

Fig. 21.1 The infrared spectrum of the reflection nebula NGC 7023 in the constellation of Cepheus showing prominent UIR features.
The *yellow curve* is the spectrum of NGC 7023 taken by the *Spitzer Space Telescope* and the UIR features are marked by the *red numbers* indicating the wavelengths of the features

carbon is tied up in organic matter in the diffuse interstellar medium. This is an astonishingly high fraction, considering that carbon can be in many forms, namely neutral atoms, ions (e.g., C^+), molecules (e.g., CO) and inorganic solids (e.g., SiC).

Not only is organic matter found in our Galaxy, it is in other galaxies as well. Astronomers know that galaxies contain significant amounts of dust as many optical images of galaxies show wide presence of dark lanes. These dark lanes are formed as a result of interstellar dust absorbing the visible star light. An example is shown in Fig. 21.2 where the galaxy M82 in the constellation of Ursa Major is seen to have dark lanes across the center of the galaxy. When M82 was observed by infrared spectroscopic observations with the *IRAS* satellite, it was found to show the aromatic signatures. The signals are so strong that it implies that there is a large quantity of organic matter in this galaxy.

One of the advantages of astronomical observations is that our knowledge is not limited to the present. Because light speed is finite, light from distant galaxies takes a long time to reach us. When we look at distant galaxies, we are also looking back in time. In the early twentieth century, the availability of telescopes of increasingly larger size led to the discovery that distant galaxies are moving away from us. This is based on the physics principle called the Doppler Effect, where light emitted by an object moving away from us will appear to shift to longer wavelengths. Through accurate measurements of distances to galaxies, Edwin Hubble (1889–1953) found that the more distant a galaxy is, the faster it is moving away from us. We now interpret this relationship to be due to the general expansion of the Universe.

Fig. 21.2 Optical image of the galaxy M82. M82 is an extremely dusty galaxy, containing a large amount of solid-state material. It is the best example of a nearby galaxy having a rich content of organic compounds. Photo credit: NOAO

Table 21.1 Redshift, distance, and time

Redshift	Receding velocity as a fraction of light speed	Present distance (in units of million light-years)	Time since light emitted (millions of years)
0.00	0.000	0	0
0.01	0.010	137	137
0.10	0.095	1,350	1,290
0.20	0.180	2,640	2,410
0.50	0.385	6,140	5,020
1.00	0.600	10,800	7,730
2.00	0.800	17,100	10,300
3.00	0.882	21,100	11,500
4.00	0.923	23,800	12,100
5.00	0.946	25,900	12,500
6.00	0.960	27,500	12,700
10.00	0.984	31,500	13,200
∞	1.000	47,500	13,700

Since the universe is expanding and more distant galaxies expand faster, the spectral lines that they emit progressively shift to redder colors as the result of the Doppler Effect. Assuming that we have a good model of how the Universe is expanding, we can relate the change in line wavelengths (the redshift parameter) to distance. This table gives some examples of the relations between redshift, galaxy receding speed compared to the speed of light, the distance of the emitting galaxy from us, and the time it took for the light to reach us

Assuming that this relationship, called the Hubble Law, is universally true, it would also allow us to use the amount of red shift to determine the distance to a galaxy. A galaxy which has a high redshift tells us that it is far away from us, and also that we are seeing that galaxy as it was some time ago. Some examples of these relationships are listed in Table 21.1. Modern telescopes of the twenty-first century can detect galaxies with redshifts as high as 6 or 7, meaning that we are seeing these galaxies as they were 13 billion years ago, not long after the creation of the Universe.

The redshift parameter is generally measured by spectroscopic observations, usually through observations of the atomic lines from the hydrogen atom. However, the same technique can be applied to the spectra of molecules such as CO so long that spectral lines of known wavelengths can be observed in the galaxy. When we perform spectroscopic observations of galaxies with large redshifts, we are seeing atoms, molecules, and solids which were made billions of years ago. The *Spitzer Space Telescope* (Chap. 9) was the first infrared telescope that was capable of performing large-scale infrared spectroscopic observations of distant galaxies. Soon after its launch, scientists found that some very distant galaxies possess the UIR emission features that are characteristics of aromatic compounds (Chap. 10). In 2007, John-David Smith of the *Steward Observatory* of the University of Arizona (now at the University of Toledo in Ohio) compiled the *Spitzer Space Telescope* observations of 59 galaxies showing the UIR features. The most distant ones have redshifts of ~2, corresponding to a distance of 10 billion light years (Table 21.1). This means that as far back as 10 billion years ago, the Universe already had manufactured complex organic compounds. Organic compounds are ubiquitous in the Universe.

Is there any relation between the organics found in the Solar System, stars, the interstellar medium, and galaxies? The only thing we know for sure is that complex organics are being made in old stars, and rapidly. Their discovery in proto-planetary nebulae shows that complex organics are being made in near vacuum conditions on very short time scales (Chap. 10). Since the lifetimes of proto-planetary nebulae are only a few thousand years, and the UIR features are not seen in their progenitors AGB stars (Chap. 4), the sudden appearance of the UIR features in proto-planetary nebulae means that these compounds must be synthesized on time scales of no more than thousands of years.

The most dramatic illustration of the ease of the formation of organic compounds by stars can be found in novae. Novae are stars that have undergone a sudden increase in visible brightness (Fig. 21.3). This is attributed to the onset of nuclear reactions on the surface of white dwarf stars after they have collected enough materials from their companion stars. Unlike a supernova which completely destroys a star, a nova only loses its accumulated outer layers and the core of the star remains unchanged.

Following this nuclear ignition is an ejection of materials from the surface in a stream of gas flowing out of a nova. When novae are observationally monitored in the infrared, it is found that a few weeks after the outburst, an excess of infrared radiation develops, signaling the formation of dust. This sudden appearance of dust can be over a period as short as 3 days. This discovery was first made by Ed Ney of the University of Minnesota in 1976. With the installation of an infrared detector on the 0.76-m telescope in St. Croix River, Ney followed Nova Vulpecuulae after its outburst on October 21,1976 with nearly daily observations. On December 23, the spectrum of the nova changed from a gas spectrum to a dust spectrum, suggesting that dust had suddenly condensed out of the gaseous ejecta.

Spectroscopic observations have shown that some novae make organic dust, displaying the UIR features characteristic of complex organics. The case of novae is perhaps the most persuasive evidence that organic matter is easily made by Nature.

On May 14, 2008, a bright optical source appeared in the nearby galaxy NGC 300. This sudden outburst resembled that of a supernova, but the output energy seemed too

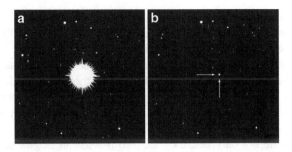

Fig. 21.3 Nova Herculis 1934.
Novae are stars that brighten over a very short period of time. The left panel shows Nova Herculis compared to its image before outburst (*right panel*). The increase in visual brightness can be a factor of a million, and typically lasts for a few weeks. It is interesting to note that this brightness decline is accompanied by an increase in ultraviolet light output and the decrease in visual brightness does not reflect an actual change in the total light output from the star

small to be a supernova. Kris Sellgren, together with her graduate student Jose Prieto, applied to the *Spitzer Space Telescope* for emergency Director's discretionary time to observe this source, now called NGC 300-OT. An infrared spectrum was obtained on August 14, 93 days after the outburst. Two prominent broad emission features at 8 and 12 μm were observed. These two broad features look exactly the same as those that we found in the spectra of proto-planetary nebulae (Chap. 10). If these features in NGC 300-OT were developed as the result of the outburst, then it suggests complex organics can be formed in a stellar explosion.

How about the organics in galaxies? Since we do not have infrared cameras with the spatial resolution needed to pin point where the organics are in galaxies, we can assume that the UIR emissions that we see in galaxies are the result of integrated emissions of organics in the interstellar medium of these galaxies. Using our own galaxy as a guide, it is quite possible that the ultimate origin of extragalactic organics can also be traced to the old stars of these galaxies.

In contrast to the galaxies, we know a great deal about the organics in the Solar System, in part because of our ability to physically examine them in the laboratory. The common wisdom has been that the Solar System organics were formed after the condensation of the Solar System, out of the gas and dust of the primordial solar nebula. This belief was shaken by the discovery of pre-solar grains (Chap. 16), the existence of which demonstrated that stellar dust particles can travel across the Galaxy and be deposited in the Solar System. They have also survived the 4.6-billion-year history of the Solar System and have remained intact for us to examine on Earth.

But how did these organic grains, which are made in stars, get to our Solar System? From the spectroscopic observations of molecules in old stars (Chap. 9), we know that these molecules are being ejected from the stars at speeds of 10–20 km/s. Considering a bullet fired from a modern rifle has speeds of about 1 km/s, these molecules are ejected at very high speed indeed. We call this phenomenon "stellar wind". By the early 1970s, it was recognized that stellar wind is a common property of old stars. By the time a star has evolved to be an AGB star (Chap. 7), winds begin to blow from the stellar surface, carrying atoms and molecules with them. These winds begin as soon as dust is formed. Light pressure from these very luminous stars is enough to drive the solid dust to high speeds. Since the molecular gas is mixed with the solids, the gas is also dragged along with it. It is through this radiation pressure mechanism that newly manufactured atomic nuclei (from nuclear reactions), molecular gas (from chemical reactions among atoms), and solids (condensed from the molecular gas) are ejected from stars.

These winds are not only fast, they are also massive in magnitude. Again from molecular spectroscopic observations, we can estimate how much mass is being removed from the stars by these winds. A typical rate is about 600 trillion tons per second. In the Milky Way Galaxy, about one star dies every year. Therefore old stars bring huge amounts of freshly made molecules and solids into interstellar space. Stellar winds from old stars are now recognized as a major source of chemical enrichment of the Galaxy. The molecular factories of AGB stars (Chap. 9) and planetary nebulae (Chap. 10) provide the raw materials from which new stars are made.

As molecules and solids are ejected by these stars into the interstellar medium, organics can accumulate in the interstellar medium over many generations of stars. Evidence for the wide presence in the diffuse (near vacuum) space between stars can be found in the detection of the 3.4-μm aliphatic feature (Chap. 17), as well as in the observations of the UIR feature in emission throughout the Galaxy. Since new stellar systems (together with their planets) are formed out of interstellar clouds, it is quite possible that the solar nebula from which our Solar System formed also inherited organic stardust. These organics are embedded in the icy comets which condensed out of the solar nebula. Bombardments of the Earth by comets may have brought a large amount of these organics to the early Earth (Chap. 18). Bombardment by km-size asteroids and comets with masses of over a billion tons probably peaked during the Hadean eon 4.5–3.8 billion years ago. Comets that have exhausted their volatile outer layers would still retain their solid nuclei made up of organic compounds. Collisions of remnants of cometary nuclei with the Earth can deliver organics to Earth. Assisted by these complex organics, life on Earth may have had a much easier time developing than if everything had to be built up from the simplest molecules.

In addition to the possibility of stardust being brought to Earth by external impacts, there is also the possibility that we may have gone to them. All the 100 billion stars in the Galaxy rotate around the center of the Galaxy. Those near the center rotate faster than those on the outside. Since the Sun is located in the outer part of the Galaxy, its circuit around the Galaxy is relatively slow, taking about 250 million years to go around the center of the Milky Way once. During this periodic voyage, the Sun must have cruised through our galactic neighborhood many times. The Milky Way galaxy is like a giant spiral, with several arms where there are concentrations of molecular and dust clouds. These clouds have masses thousands to millions of times the mass of the Sun, and have sizes as large as millions of times the size of the solar system. When the Sun passes through such clouds, the dust in the clouds must have rained on the Earth at much higher rates than normal.

Even with the benefit of externally delivered organic compounds, there is still a long way for life to have developed on Earth. Scientists believe that life is a chemical phenomenon, and the problem of the origin of life is to find a set of chemical pathways that can transform the basic ingredients into self-organizing entities, the simplest forms of life. That life began from abiotic materials is known as the Oparin-Haldane hypothesis (Chap. 1), which is the premise we started from on the question of the origin of life.

In order for a living system to be created from non-living matter, several building blocks of life are needed: (1) amino acids for building proteins; (2) phosphate for the backbone of nucleic acids; (3) purine and pyrimidine bases for the genetic code of DNA and RNA; (4) sugars for the nucleic acids; and (5) fatty acids for cell membranes. In life forms as we know them, proteins serve the role of biological catalysts, and nucleic acids such as DNA and RNA carry the genetic information from one generation to the next. Astronomical observations have already detected simple forms of sugar (Chap. 17) and searches for the basic structures of nuclear bases such as purines and pyrimidines are ongoing.

Proto-cells are made possible by the formation of membranes through the synthesis of lipids or fatty acids. In present life, the lipids used in the formation of cell membranes in archaea are different from those in bacteria and eukarya, and the pathways to their synthesis are different. While lipids in bacteria consist of ester bonds between glycerol and fatty acids, lipids in archaea consist of chains of 20 or 40 carbon atoms. As for possible precursors of lipids, there are long-chain molecules made up of hydrogen, carbon, and nitrogen (called cyanopolyynes) that have been extensively detected in old carbon stars. Straight carbon chains such as C_3 and C_5 are also known to be present in the circumstellar and interstellar media. Although the likelihood of astronomical detection of fatty acids and amphilphilic molecules that make up cell membranes is remote at present, long-carbon chains have been observed and these could be combined to form longer chains.

In order for life to occur, we need ingredients more than just hydrogen and carbon. Nitrogen is an essential element in biochemistry. Hydrogen cyanide (HCN) is a molecule commonly found in interstellar space. Consequently, there is a lot of interest in HCN polymers, which have been shown in the laboratory to yield amino acids, purines, and other precursor constituents of RNA. In Chap. 15 we talked about the possibility that HCN polymers are among the types of organic matter that can be made naturally in space.

Next, there is the question of an environment for life to develop in. Since liquid water seems to be an essential element for life, scientists envision a "primordial soup" from which life arose. More recently, the idea of submarine hydrothermal vents as the nursing ground for life was proposed. Unfortunately the Earth has changed a great deal since its early days. The surface topography, atmospheric composition, ocean acidity, etc. have all evolved and we are left with little direct evidence of the conditions under which life began. The earliest evidence for life comes from the apatite grains in southwestern Greenland, where the existence of microorganisms can be inferred. From the age of the rock, one can therefore conclude that life began at least 3.86 billion years ago. There have also been suggestions that there are microfossils in Australian Apex sediments, inferring evidence for life at 3.5 billion years ago. If we accept this early rise of life, then we are left with a several hundred-million-year gap between the formation of prebiotic molecules and the creation of life. Our scientific understanding of what transpired during this gap is very limited. Beyond chemical ingredients and chemical reactions, are we missing any other essential elements?

Our present situation in the quest for an understanding of the origin of life reminds me of astronomers' attempt to understand the structure of stars about 100 years ago. At that time, physicists and astronomers already possessed a good understanding of gravitation and thermal physics. Some rudimentary understanding of atomic physics was also in place. In fact, many physicists believed that everything about the physical world had already been discovered. A star was known to be a gaseous body and it seemed all the elements were there to have a complete physical description of its structure. By the early twentieth century, the British astrophysicist Arthur Eddington (1882–1944) had made sophisticated mathematical models of the structure of stars, taking into account how the gaseous body is held together by gravitation and prevented from collapsing upon itself by thermal gas pressure. By treating the absorption of light by atoms, one could calculate how the density and temperature of a star change from the inside to the outside. The output of light from the stellar surface can also be calculated. All very good except one thing: a star like the Sun would only shine for tens of millions of years, not billions of years as demanded by geological evidence.

The evolution of astronomy

Astronomy is an ancient discipline which can be traced back over 4,000 years. In the beginning, the practice of astronomy concentrated on tracing the movements of stars, the Moon, the stars, and the planets. Mapping and cataloging of stars and their positions were also important in our understanding of the heavens. The first change in these practices began with the Greeks applying the newly developed tools of geometry to construct models of the Universe. The discipline of mathematical astronomy was brought to a new height by Ptolemy and continued by Copernicus. The realization that the movements of planets are the result of physical forces by Kepler, Galileo, and Newton introduced physics into the discipline of astronomy. The high point of astrophysics was in the first half of the twentieth century when atomic and nuclear physics brought us a quantitative understanding of gaseous nebulae (Chap. 8) and stellar structure, respectively. From the 1970s on, the discovery of interstellar molecules brought in the discipline of astrochemistry (Chap. 9). Now at the beginning of the twenty-first century, we see the emergence of astrobiology. Whereas in the past our sample of biology has been limited to the Earth, the potential now exists to discover life elsewhere, in our Solar System or even beyond.

This was an insurmountable problem no matter how clever one was at that time. It turned out that one element was missing: nuclear physics. In 1896, Henri Becquerel (1852–1908), Marie Curie (1867–1934) and Pierre Curie (1859–1906) discovered radioactivity. In 1910, Ernest Rutherford (1871–1937) found that atoms have very small nuclei where most of their mass resides. In 1932, James Chadwick (1891–1974) discovered the neutron, a neutral particle inside the atomic nucleus. Most importantly, the discovery of nuclear decay showed that chemical elements can naturally change from one element to another. A neutron inside an atomic nucleus can naturally decay into a proton, therefore changing the atom into the next element in the periodic table. Nuclear reactions can go two different ways: either a heavy nucleus can breakdown to lighter nuclei (a process called nuclear fission), or lighter nuclei combining to form a heavier one (a process called nuclear fusion). Just as molecules can change from one to another through chemical reactions, elements can change from one to another through nuclear reactions. This change is accompanied by a release of energy, a source of energy never envisioned before.

Until the properties of atomic nuclei were known and the nuclear reactions of fusion were understood, all attempts to understand stars were futile. Are we in a similar stage on the question of the origin of life?

It has been said that the three most important questions in science are: (1) the origin of the chemical elements; (2) the origin of life; and (3) the origin of the Universe. The question of the origin of chemical elements was solved in the 1950s as the result of development in nuclear physics (Chap. 4). The discovery of cosmic background radiation in the 1960s provided many clues to the early days of the Universe, which most astronomers today believe to have arisen from a Big Bang. However, since the Universe was opaque in its first 400,000 years, it is impossible for us to observe the early moments of the Universe no matter how powerful telescopes we can deploy. Although many theories have been constructed, we may never know for sure how the Universe got started. There is also the problem that the Universe is unique. There are no other universes that we can use for comparison. Our faith in scientific understanding is based on samples. We believe we have a reasonably good understanding of the structure and evolution of stars because there are many stars in the Galaxy for us to study. We think we understand the process of evolution because we have many living species on Earth. Our theoretical understanding of the process of evolution is based on a synthesis of the behavior of many species. Any patterns that we observe in one species can be used to predict the properties of another species. Confirmations of such predictions are powerful endorsements of the theory. Unfortunately, any theory that we put together for our Universe cannot be used to predict how other universes got started. It is not to say that we cannot learn more and more about the early days of our Universe; it just means that whatever theory we develop has limited predictive power.

The question of the origin of life is different. We have a clear example of life on Earth and we have made significant progress in learning how life developed on Earth. Admittedly, there are still many gaps in our knowledge. Our ability to send probes to Mars and planetary satellites such as Titan gives us hope that in the not too distant future we may discover life beyond the Earth. The discoveries of life forms under extreme hardship conditions on Earth, in the driest deserts to hot undersea thermal vents, give us hope that life may also exist on Mars, where conditions are not that different from some areas on Earth. If we push the limits a little further, it is not impossible that planetary satellites such as Europa and Titan can also harbor life on or under their surfaces. Planetary exploration missions in the next 100 years will carry out conclusive tests of these possibilities. Consider the fact that the human race has been wondering about the origin of life for several thousand years, and the answers may be within reach in the near future.

When life is detected elsewhere, analyses of the chemical makeup of other forms of life will greatly help us understand how life got started from lifeless biochemical molecules. On the other side, the analysis of stardust, samples returned from asteroids and comets, will tell us how complicated extraterrestrial biochemical molecules can be. With battles fought from both ends, we may finally fill in the gaps of our present knowledge of the origin of life.

At this juncture of the twenty-first century, scientists have never been as optimistic as now on solving the problem of the origin of life. During our pursuit for an understanding of our origins, we have established that the physical laws that govern events on Earth are the same ones that govern the planets and the stars. The chemical makeup of matter on Earth can be traced to elements made in stars. Now we are pondering the possibility that biology on Earth may have been influenced by delivery of tiny particles from stars faraway. In spite of the large distance separating us, our lives on Earth are very much tied to the stars.

A brief summary of this chapter
Existence of organic compounds in the Universe and the possibility that they were delivered to Earth.

Key words and concepts in this chapter
- Organics in galaxies
- 10 billion years of organic compounds in the universe
- Synthesis of organics by novae
- Stellar winds from old stars and the spread of organics throughout the Galaxy
- Chemical pathways leading to life
- The primordial soup
- Conditions for the development of life

Questions to think about
1. Do you think we will ever find the answer to the question of the origin of life? Is the answer beyond the realm of science?

2. Similarly, do you think that the question of the origin of the universe can be answered? Our current thinking of the origin of the universe is based on a number of assumptions, among which are "the laws of physics are the same everywhere and at any time". Is there a possibility that this assumption may break down?

3. Aristotelian cosmology is based on the principle that the Universe is perfect. Our present cosmology is based on the underlying principle that physical laws are universal in both space and time. The Aristotelian principle was broken by the invention of the telescope, showing that celestial objects are imperfect, and the derivation of the Kepler's laws of motion where planets are shown to be moving in non-circular orbits. What experiments are needed to test the universality of physical laws?

4. Is there a limit to how far and how much we can observe the universe? The construction of larger and larger telescopes has extended the distance limit on how far we can study an astronomical object. Can this be continued indefinitely?

Chapter 22
Our Place in the Universe

In the previous chapters, we have gone through a long list of recent astronomical discoveries that point to the possible extraterrestrial influence on biological development on Earth. In particular, we have discussed how easy it is for stars to make complex organic compounds and how large quantities of very complex organic compounds are made over very short time scales under extremely low density conditions. How stars manage to do this is beyond our present theoretical understanding but from an observational point of view, there is no doubt that this is happening. Not only are stars able to manufacture organic compounds, they are also able to eject and distribute them all over the Galaxy, and there is evidence that some of these organics were embedded in the early solar nebula. Evidence of these organics is found in all kinds of Solar System objects, including comets, asteroids, meteorites, interplanetary dust particles, and planetary satellites. External bombardment brought these organics to the early Earth and might have influenced the development of life on Earth. If this stellar-Earth connection is real, then a similar process could easily have occurred on other planets in millions of other solar systems in our Galaxy. Given the appropriate environment, life, most likely in a form different from our own, can flourish. It is possible that life is common in the Galaxy, and we are far from unique, and certainly not a chosen people.

In order to put this in perspective, we should take a look at the recent history of scientific development regarding our evolving perspectives of Earth and ourselves in the context of the Universe. For thousands of years, we thought we occupied the center of the Universe and the Sun, the Moon, the planets and the stars all revolved around us. This concept of self-importance was shattered by the heliocentric model of Copernicus 500 years ago. The Earth was demoted to the status of an ordinary planet, no different from Mercury, Venus, Mars, Jupiter and Saturn, the five planets known at the time. With the invention and deployment of telescopes as a tool for celestial observations, we discovered new planets such as Uranus and Neptune, as well as their many satellites. Some of the satellites, e.g. Titan of Saturn, have atmospheres and oceans and are in many aspects similar to the Earth.

S. Kwok, *Stardust*, Astronomers' Universe, DOI 10.1007/978-3-642-32802-2_22,
© Springer-Verlag Berlin Heidelberg 2013

However, the Sun was still thought to be unique. It is clearly much brighter than any other celestial object and is in a class of its own. Our Earth is blessed with the proximity of the Sun, which seemed to make us special. In the time of Aristotle and Ptolemy, stars were believed to lie on the surface of the celestial sphere and they were believed to be far away but all at the same distance from us. With the invention of telescopes, scientists gradually came to the conclusion that stars have different distances from us, but the exact values of these distances were not known.

In 1838, Friedrich Wilhelm Bessel made the first accurate measurement of the distance to the nearby star 61 Cygni using the method of parallax. The method of parallax works in the following way. As the Earth revolves around the Sun, our perspective to a star changes. If we compare the position of a star when we are on one side of the Earth's orbit (say, during winter), the stellar position would be different if it is observed when we are on the other side (say, during summer). Since stars are at different distances from us, a nearby star will shift its position more than a distant one. So if we compare the patterns of stars in the sky from one season to another, we should be able to detect a change in pattern as distant stars will not shift but a nearby member will. However, for a long time no such shift was detected and this in fact was used to argue against the heliocentric hypothesis of Copernicus. If the heliocentric theory is correct, then stars must be very far away.

A successful parallax measurement requires tremendous accuracy in angular measurements. When Bessel was able to make this measurement, it was the first time that the distances to stars could actually be measured. The distance to the star 61 Cygni was found to be 10.3 light years. Later observations showed that the nearest star to the Sun is Alpha Centauri, which is 4 light years away. If we adjust for the decrease in apparent brightness with increasing distance, it turns out that if we placed the Sun at a distance of light years away, it would look no different from other stars. This proves without doubt that the Sun is a star, our nearest star.

Consequently, the Sun also lost its special significance. It is no different from any one of the thousands of stars in the sky. Instead of being an object in a class of its own, it is one of many stars. Yet, astronomers held the view that the Sun was at the center of the known Universe. At the beginning of the twentieth century, the Dutch astronomer Jacobus Kapteyn (1851–1922) established that the Milky Way galaxy is large—in his estimation measuring 30,000 light years in diameter and 6,000 light years in thickness. He concluded that the Sun is one of the millions of stars in the Galaxy and is located right in its center. The Sun, our parent star, substituted the Earth as the center of the Universe.

In the following 20 years, astronomers also began to suspect that there are other nebulous objects (which we now call galaxies) that could be collection of stars located far away from us. The watershed event was the debate between Harlow Shapley (1885–1972) and Heber D. Curtis (1872–1942) on April 26, 1920 that took place at the meeting of the National Academy of Science in Washington, D.C. Shapley believed that the nebulae are gas clouds within our large galaxy in which the Sun is located far from its center. Curtis, on the other hand, argued that there are other galaxies beyond our own and the Sun is at the center of our galaxy. In fact, both were wrong. Our Milky Way galaxy is indeed large and the Sun is an insignificant outer member of this galaxy, but there are many other galaxies like our own. Through better measurements of distances, we now know that there are hundreds of billions of galaxies in the Universe, each containing hundreds of billions of stars like the Sun. The discovery of interstellar dust led to our realization that it obscures much of the Milky Way from our view and our picture of the Milky Way is far from complete. The Sun, instead of being located at the center, in fact, is near the edge of the Milky Way. We in fact hold no special place in the Universe and the Sun and the Earth are very insignificant and common members of the galactic family.

In addition to realizing our diminishing place in space, we are also shrinking in the domain of time. Our concept of time is based on our own personal perception of transitory events and our memories of the past. This can be expanded to historical time when written records are taken into account. The oldest reliable written history is probably by the Chinese, whose written historical record goes back to the Xia Dynasty (~2000 BC). Beyond that, we have no sense of time. Aristotle argued that time should not have a beginning, but should be cyclical as the circular motions of stars.

The rise of Christianity meant that the Biblical concept of Creation demanded a beginning. Various biblical scholars from St. Basil the Great (fourth century), St. Augustine (fifth century), Kepler and Newton gave estimates of the time of creation to be ~4000 B.C. In 1650, the Archbishop James Ussher of Ireland (1581–1656), calculating from biblical genealogy in the Old Testament, concluded that the Creation occurred in 4004 B.C., putting the Earth at less than 6,000 years old at that time (or 6019 years as of 2012).

The realization that the shaping of the Earth was done through slow processes such as erosion and deposition over a much longer time scale did not happen until the nineteenth century. The view that the Earth is constantly changing and has not been the same all along was gradually accepted. This view was first promoted by the Scottish geologist Charles Lyell (1797–1875), who maintained that all geological and biological changes were the results of natural causes throughout the Earth's history. Several different techniques have been developed to give a scientific estimate of the age of the Earth, including the accumulation of sediments in water bodies, salinity of the oceans, and fluorine in bones. We know that the rain falls, the fresh water leaches and carries different chemical elements from the rocks to the oceans. Since the oceans have no outlet, dissolved salts accumulated over time and the oceans became salty. Assuming that there is no removal mechanism (which we now know to be untrue), the age of the Earth can be estimated from the accumulation rate. From estimates based on the salinity of the oceans and the rate that salts are carried by rivers, Irish geologist John Joly (1857–1933) extended the estimate for the age of the Earth to 80–90 million years. Around the same time, William Thomson (later became Lord Kelvin, 1824–1907) arrived at the age of 20–40 million years by assuming that the Earth was originally molten and using the cooling rate of the Earth to calculate the amount of time that it needed to cool. In 1856, Hermann von Helmholtz, assuming that the Sun is a ball of gas held together through a balance of gravitational self-attraction and thermal pressure of the gas, concluded that the solar luminosity is the result of release of energy through gradual contraction of the Sun. This leads to a lifetime of about 22 million years. Using similar arguments with estimates of the amount of heat possessed by the Sun, Lord Kelvin came to an estimate of 20–400 million years in 1864. There was no secret that he favored the lower end of the values. In 1897, Lord Kelvin revised the estimate down to 20–40 million years.

Since Lord Kelvin was such a towering figure in science, even great scientists of the time, including Charles Lyell and Charles Darwin, conceded to his views. It is interesting to note that Darwin retracted his original estimates of 300 million years for the age of the Earth (derived from erosion and weathering of rock) in the later editions of *The Origins of Species*. The influence of Lord Kelvin can be attributed to his strong and forceful personality and the belief that the Earth is less than 100 million years old remained fashionable through the late nineteenth century.

Although Lord Kelvin's arguments were rigorous and elegant, they were unfortunately incorrect. Nuclear physics was not known at that time and he was therefore unaware of the other possibilities for energy generation in the Sun. Progress in nuclear physics led Arthur Eddington to speculate nuclear fusion as a possible source of energy. This was eventually confirmed by the work of German-American physicist Hans Bethe (1906–2005) of Cornell University in 1939.

An accurate way of determining the age of Earth came with the technique of radioactive dating. Through the recognition that lead is a decay product of uranium, the U.S. chemist Bertram Boltwood (1870–1927) came up with an age of 400–2000 million years in 1907 through the measured ratio of lead and uranium in rocks. With a similar method, the British geologist Arthur Holmes (1890–1965) in 1927 increased the estimate to 1.6–3 billion years. This was further increased to 3.4 billion years by Ernest Rutherford (1871–1937) in 1929. Using lead isotopes, Arthur Holmes arrived at a value of 3.35 billion years in 1947.

Refinements of the radioactive dating method led to the identification of the oldest rocks (which are in Greenland) as 3.8 billion years old. By extrapolating backward in time to an age where there should have been no lead produced by radioactive decay, our current estimate of the age of the Earth is 4.6 billion years.

Since the Earth condensed out of the solar nebula, the Solar System must be at least as old as the Earth. Carbonaceous chondrites contain a number of small inclusions of unusual high-temperature minerals and these must have condensed directly from the solar nebula. The dating of these inclusions gives an age of 4.565–4.568 billion years for the Solar System.

It is interesting to note how recent our modern view of the age of the Earth is. Our current estimate was arrived at just half a century ago. In a short period of 300 years (from 1650 to 1950), our view of the age of the Earth has changed from several thousand to several billion years. Our existence as human beings, having come on the Earth only about a hundred thousand years ago, represents a small fraction of the history of the Earth.

Another long held belief is that the Earth is an isolated place undisturbed by other events. For a long time scientists believed that craters on Earth and similar craters on the Moon were volcanic in origin. As late as the mid-twentieth century, it was still very popular in the Soviet Union to believe that the lunar craters were created by volcanic eruptions. This debate formally ended with the Apollo landings on the Moon, where the lunar rock samples were found to contain glass formed by impacts. May be we should not be so hard on the Soviets because during most of the twentieth century, many geologists in the U.S. maintained that the magnificent crater in Arizona was the product of steam explosion (Chap. 2).

After the acceptance of the extraterrestrial origin of meteorites in the mid-nineteenth century, the scientists who study meteorites formed a close community. They go to the field to find new meteorites, analyze them in the laboratory for their chemical composition, and speculate on their origin within the Solar System. Never would they imagine that these rocks contained stuff coming from outside of the Solar System, let alone a star. Until the discovery of pre-solar grains (Chap. 16), they never felt the need to talk to astronomers. Once they started doing that, they found that astronomers who study nucleosynthesis in stars could help them interpret their data, and they in turn could use the meteorite samples to test the stellar models of astronomers.

There are other examples of the conservatism of the scientific community. The important concept of continental drift, where major geological features such as mountains were made through the movement and interaction of large rigid plates on the Earth's crust, was not accepted until the 1960s. Our world is not constant and land is in fact continuously shifting under our feet. Back in 1660, before the availability of accurate maps, Francis Bacon (1561–1626) already noticed that the new continent of South America fits the shore line of Africa like a jigsaw puzzle. Could the continents have moved over geological times? In 1912, a German meteorologist, Alfred Wegener (1880–1930), noticed that fossils from different parts of the world are remarkably similar. He went on to propose that our present continents originated from a one piece supercontinent called Pangaea that broke up 200 million years ago. He did not receive much favorable response from the geology community and died in 1930 before his correct hypothesis was widely acknowledged much later.

Geologists are not the only stubborn ones; astronomers can be too. The delay in the acceptance of the aromatic origin of the UIR features was in part due to their lack of familiarity with organic chemistry and laboratory spectroscopy, and their preconceived notion that nothing that complicated could exist in space. Now with overwhelming evidence for the existence of interstellar molecules and very strong evidence for the existence of complex organic compounds, the defense line has retreated to the impossibility of life in space. Many prebiotic molecules, including amino acids, have been found in meteorites, demonstrating that prebiotic chemistry is active beyond the Earth. Will viruses and bacteria be found in the Solar System in future space exploration and sample return missions?

While modern scientists believe that we have outgrown Aristotle, in fact his shadows are still with us. Many astronomers hang on to the idea that the universe is pure and simple. Although almost everyone has accepted the idea that gaseous matter is the constituents of stars, and even interstellar matter, many still resist the ubiquitous nature of solid state materials, and the presence of minerals in stars and galaxies.

This can be attributed in part to the fact that infrared astronomy is a relatively young technique. Infrared spectroscopy, which forms the basis for the detection of solid matter in space, is even newer. But I suspect that deeper in the astronomers' psyche is the distinction between the Heavens and Earth. Astronomers place the study of stars, galaxies, and the early universe on a higher plateau, and look down upon the study of the Solar System, which they regard as dirty and complicated. Aristotle's sub-lunary sphere referred to the Earth and super lunary sphere contained the planets and the fixed stars. While Aristotle regarded the Earth as muddy and changeable, this description is now used to apply to the Solar System. Astronomers have shifted their domain of study to stars and galaxies and left the Solar System to the discipline of planetary science. In this book, I have tried to convey the message that the Heavens and Earth, or more precisely, stars and the Solar System, may be more connected than we used to think.

Aristotle and his contemporaries regarded the stars and the planets as perfect because they lacked the ability to know better. Now we have landed on the Moon, Mars, and Titan, and have flown near Mercury, Venus, asteroids, and comets. From these close encounters, we know that these heavenly bodies are complex and even Earth-like in some ways. The planets are just points of light to the Greeks, the same way that astronomers now look at distant galaxies. Does our pursuit and passion for distant points of light and disdain for the Solar System come from a subconscious belief in Astristotle?

Although the telescopes at astronomers' disposal are now much bigger and more powerful than ever, interestingly the observational techniques have changed little. Many modern astronomers still devote their energy to taking pictures and measuring brightness (imaging and photometry respectively in astronomical jargon), the same way that astronomers did a hundred years ago. The only differences are the targets and the use of charge coupled device (CCD) cameras instead of photographic plates. Astronomers now chase after faint, distant galaxies the same ways that astronomers a hundred years ago chased after faint and distant stars. The technique of spectroscopy, which has been responsible for our physical understanding of the universe, plays second fiddle to imaging and photometry in terms of popularity.

Our supposition of the stellar-Earth connection is through external delivery. While the theory of external delivery was never taken seriously in the twentieth century, opinion has almost completely turned around now. This was achieved not by a single individual or even a single community, but by collective efforts of astronomers, planetary scientists, geologists, and laboratory chemists. This way of working across traditional disciplinary boundaries is foreign to many scientists, as scientific training has become highly specialized and scientists are mostly only comfortable talking to their own kind.

It is commonly believed that science progresses by the accumulation of facts and through logical arguments. While this is true to some extent, it ignores the human element. The formulation of scientific thoughts and theories is a creative process, no different from drawing a painting or composing a piece of music. We also have to recognize that it is difficult to make conceptual changes in one's mind.

The scientific establishment, often with the most to lose from radical new ideas, has been repeatedly shown to be obstinate. Breakthroughs are often made by an outsider, with no vested interest in the status quo. It is often difficult to sever emotional attachments to old ideas, discard familiar tools, argue with superiors and people of authority, or risk one's position and security, in order to fight for the elusive truth. As a scientist today, it is much easier to go with the flow and ride the bandwagon, especially when research grants and tenure are at stake. To offer a new idea and challenge the establishment takes courage.

Until about the 1960s, the power of conducting research was concentrated in a small circle of elderly figures. This was true in England, the United States, and Europe. Funding was tightly controlled through research institutes, in many cases in the hands of one man. Only during the post-*Sputnik* era when the U.S. government decided to inject massive funding into science did this practice begin to change. The large influx of federal funds to universities broke the old barrier and allowed junior members in U.S. universities to have direct access to research funds through competitive applications to federal agencies such as the National Science Foundation, the Department of Energy, the National Institute of Health, and NASA. The democratic distribution of funds based on the principle of merit is probably the single most important factor in the knowledge explosion in the last half of the twentieth century. This U.S. system was followed by other countries such as Canada and the U.K., and later by Germany, France and other western European countries. Countries that were slow in adopting the open grant competition system, which include Russia, Japan and other Asian countries, lag behind in research competitiveness.

As science becomes more specialized, it is very easy to lose the global view. As we have seen in this book, modern advances in our understanding of our origins require a broad understanding of astronomy, physics, chemistry, biology, geology, and planetary science. When one does not understand something, it is much easier to defend one's own turf and stand by tradition. It is particularly easy to accuse an "amateur" of his lack of in-depth expertise and dismiss his new views. While many scientists and academics just blindly follow the bandwagon and play the game by the rules of the day, the more ingenious ones have learned to get around the system by maintaining an acceptable research program in order to sustain funding, and at the same time explore new ideas in their "spare time". Such practices are grossly inefficient. However, we have to sympathize with the scientists as they are under tremendous pressure to generate results in main stream subjects, flooding the journals with papers of sometimes marginal results. The other side of the argument is that this world is full of crackpots. How is one to separate the truly imaginative and the fools? This is a fine line indeed.

While the scientific debates of the past were primarily between individuals, with some intrusions from the Church, modern scientific debates operate in much more complicated surroundings. Corporate interest and political influence often come into play, with global warming and climate change being prime examples. The days of "gentlemen scientists"—people of independent means engaged in research—are long gone. Now every scientist is on the payroll of some organization, be it university, government agency, non-profit organization, industry, or corporation. Research is expensive. In the twenty-first century, ground-based telescopes cost tens to hundreds of millions of dollars, and in the case of the *Thirty Meter Telescope* (*TMT*, a project by a US-led consortium) or the *Extremely Large Telescope* (*ELT*, a European project), over 1 billion US dollars in each case. The ground-based radio telescope the *Atacama Large Millimeter/submillimeter Array* (*ALMA*) currently under construction has cost over $2 billion US, and the planned *Square Kilometer Array* (*SKA*) is likely to be even more expensive. Space-based telescopes are particularly costly, routinely costing hundreds of millions or even billions of dollars (the *James Webb Space Telescope* is estimated to cost $8 billion US).

It is an established reality that frontier research can no longer be done with a beaker and a few test tubes in the basement, but requires substantial funding from the government. This implies proposals and peer reviews. Access to a good instrument, e.g. a large ground-based telescope or a satellite in orbit, requires one to submit to reviews. Unconventional ideas are simply not funded or granted observing time. A young scientist will never get anywhere if he is not a "team player". Is such a system conducive to innovation and the development of science?

Remarkably, in spite of all these handicaps, science managed to thrive and grow. Small numbers of young men and women are still able to buck the trend, and propose new theories that depart from conventional wisdom. The stories told in this book are the stories of mavericks. Some live long enough to see their ideas vindicated. Some unfortunately die before their contributions are recognized.

On the whole, I am optimistic about the progress of science. Truth always eventually prevails, although the time it takes can be long and some of the pioneers may not live to see their ideas acknowledged or recognized and their reputation vindicated. In spite of the few whose motivations are dominated by the hunger for power or blinded by ambition, most scientists are still driven by curiosity and the search for truth. I have no doubt that the answers to the question of the origin of life will come. This answer may draw upon the works that are reported in this book, or may have nothing to do with any of it. No matter how it turns out, it has been an amazing journey.

Key words and concepts in this chapter

- The stellar-Earth connection
- Earth as a planet
- Sun as a star
- The spatial scale of the Universe
- The age of the Earth
- The extraterrestrial origin of meteorites
- Sociology of science
- The imperfection of the heavens
- An interdisciplinary approach to solve the problem of origin of life

Questions to think about

1. The process of science is based on hypothesis-testing experiments. Over history, many very popular concepts of science have turned out to be wrong. These include "ether", a substance which is supposed to be permeating throughout the universe; "vitality", the essence of life; "phlogiston", the element of combustion, and many others. Are there concepts currently popular in science that could turn out to be fictitious in the future?

2. Astronomy has been referred to as a perfect example of a useless science. Why are governments willing to spend billions of dollars to build astronomical facilities?

3. If you have the power to change the system of how science is done, what form will it take? What role should the government play in this scheme?

4. In 2012, the US Congress debated whether the funding of science should be left to the private sector, as is the case for the technology sector. Do you think that would be a good idea?

A Pilgrimage to the Stars

As I step off the plane into the open air terminal building of the Honolulu airport, I am greeted by the warm tropical air and the smell of the sea. Around me are hundreds of visitors from all over the world, seeking a few days of relaxing holidays in the sun. I am used to the tourist crowd as I have been on this flight dozens of times. I have also gotten used to the puzzled glances from my fellow passengers, wondering what this crazy guy is doing carrying a heavy winter coat in this tropical paradise. Indeed I am the odd man out in this crowd as I am not heading to the beach, but instead up to a dormant volcano 4,200 m (14,000 ft) above sea level on the Big Island of Hawaii. For the place that I am going, shorts and swimming suits will not do, for I will need all the warm protection that I can get in the subzero temperature on the summit of Mauna Kea.

Mauna Kea, meaning "white mountain" in Hawaiian for its snow-capped summit, is a sacred site of the native Hawaiians, being the home of the snow goddess Poliahu. My frequent pilgrimage to this mountain is not for religious reasons, but because Mauna Kea is a holy site of modern astronomy. It is the most popular location in the northern hemisphere to place large telescopes for astronomical research. A tropical inversion cloud layer forms around the mountain at about 600 m (2,000 ft), meaning that the summit is almost always above the clouds, resulting in endless clear skies above. Nations all over the world are clamoring for the limited amount of real estate on the summit to house their most advanced telescopes. As of 2012, countries represented on the mountain include the U.S.A., U.K., Canada, France, Holland, Japan, Australia, Argentina, Brazil, Chile, and Taiwan.

A journey to the mountain begins at either the towns of Hilo on the east coast or the popular resort town of Kona on the western shore. Most observatories maintain a sea-level office; for example, the Gemini Telescopes and the Subaru Telescope are based in Hilo, whereas the Canada-France-Hawaii Telescope and the Keck Telescopes are based in Waimea. After a taxi ride from the airport, I pick up the keys for a four-wheel drive vehicle from the observatory and head up the mountain. A partly paved Saddle Road winds through the Parker range, one of the largest cattle ranges in the United States. The pasture is green and lush as rainfall is plentiful. My car moves in and out of mist and fog as I gradually climb through the inversion cloud that I mentioned previously.

Finally I emerge from the clouds and make my first stop at Hale Pohaku (Hawaiian for "stone house"), a mid-level (2,800 m or 9,000 ft) facility for the observatories. Here is where I usually spend my first night trying to acclimatize to the high altitudes of the summit. Over the years, Hale Pohaku has greatly expanded and now provides living quarters for over 70 visiting astronomers and local technicians. It also has a cafeteria, serving meals to the residents of diverse nationalities and varying tastes. No alcohol is served, although I have heard rumors of the French bringing their own wines in secret. The food is American-Hawaiian, and, shall I say, not gourmet quality. I had always hoped that when the Japanese built the Subaru Telescope, they would demand that sushi be served, but somehow this never materialized.

Dinner is early in Hale Pohaku, starting at 4 pm in order to allow enough time for the astronomers to head up the mountain and be ready to observe at dusk. Since I am not observing on this first night, I relax on the sun deck chatting with fellow astronomers who have been here for a few days, trying to catch the latest stories on instrument performance and upcoming weather conditions. Too many times, I find out that the instrument that I am expecting to use is having problems with a failed component, requiring me to make new plans to adjust to the current situation. Sometimes, wind and snow on the summit intervene, creating endless anxiety that my precious observing time may go to waste.

After dinner is time for the library. Here my students and I go over the star charts and observing plans, scheduling every detail of the coming nights down to the minute. This is necessary because at the summit of 4,200 m, our mental capacity is not functioning at optimal levels, and one does not want to make errors under the influence of the lack of oxygen.

Because of the altitude, the atmospheric pressure on the summit is only 60 % of that at the sea level, and the decreased supply of oxygen to our lungs can create acute mountain sickness including headaches, drowsiness, nausea, and worst of all for a scientist, impaired judgment. Since the cost of observing time on a telescope is valued at tens of thousands of U.S. dollars per night, one does not want to make any mistakes or waste a single minute. Thorough preparation is the key to a successful observing run.

The other problem that we have to face is time change. Not only does one have to deal with jet lag due to time difference between our place of origin and that of Hawaii, there is also the problem of changing from a day to a night schedule. Usually I try to stay up as late as I can on my first night on the mountain, going to bed at 3 or 4 am in an attempt to adjust.

After a few hours of rest and breakfast around noon, I head up to the summit in order to check out the conditions of the telescope and the instrument. The road between Hale Pohaku and the summit is mostly unpaved, steep, and winding. The landscape is volcanic rocks on cinder cones and lava plateaus. There is a total absence of vegetation, and this barren land cannot be more different from the palm trees and sandy beaches below. The road can be hazardous at times, especially during winter months when the icy conditions can be deadly.

Once on the summit, we are greeted with the most spectacular sight of white telescope domes against the background of infinite blue skies. Among the largest are the two 10-m Keck Telescopes, the 8.3 m Subaru Telescope,

(continued)

Fig. 22.1 The *Gemini North Telescope* under a starry sky.
The 8-m *Gemini North Telescope* is located on the summit of Mauna Kea. Its twin *Gemini South Telescope* is located in Chile. Photo credit: Gemini Observatory/AURA

Fig. 22.2 Sunset on Mauna Kea.
The telescopes from near to far are the *Gemini North Telescope*, the *United Kingdom Infrared Telescope* and the *University of Hawaii Telescope*. Photo credit: Gemini Observatory/AURA

and the 8 m Gemini Telescope (Figs. 22.1 and 22.2). Having introduced myself to the day crew of engineers and technicians, I confirm with them what I need for the night and that things are indeed working. In spite of the enormous sizes of the domes, the operation room is usually small and cramped. I put down my laptop computer, spread out my notes and charts, and store my drinks and night time meal in the fridge. If the telescope operator is on hand, I acquaint myself with him or her, as the operator is the astronomer's most valuable friend and partner during the night.

After returning to Hale Pohaku for a short dinner, I head up to the summit again, turn on the instruments, fire up the computer, and sit down to wait for sun down. This often is the most anxious moment because we will find out whether the instruments are working up to our expectations and whether the computer is running without any glitches. If any problem develops, we put on our boots, gloves, and heavy coats and leave our heated control room for the naturally cold dome above. As the instruments are often hung on the tele-scope high above the ground, we use a "cherry picker" to reach for the boxes. Trying to fix delicate things in such clumsy clothing (and often in darkness) is not the easiest job. If we are lucky, the problems are not major and observing can start. As the first few hours go by, we settle down to the routine of observing and can begin to relax.

People relax in different ways on the telescope. In the old days, one had to sit in front of the console to constantly guide the telescope to ensure it was not drifting from the target. In recent years, much of this tedious work has been automated and the computer actually does the guiding. Many of us still carefully watch the screen for any signs of instrument malfunctioning or abnormal weather conditions. However, there is always some time in between exposures when there is not much one can do. Some astronomers read a novel, some listen to music, some take a nap, and some pace back and forth anxiously waiting for the outcome of the observations. Will we detect any-thing? Are we going to discover a new phenomenon? Is my theory going to be confirmed? After the exposures are finished, we take a quick look at the raw, unprocessed data in the hope of getting a glimpse of what we are looking for. Disappointment will mean an adjustment of plans. Maybe we should expose longer, maybe look at a different object. A new discovery will lead to excited cheers, hugs, and sometimes tears.

Astronomers sometimes behave strangely when they are on telescopes. I know an elderly professor who looks solemn and dignified on campus but will curse and swear endlessly while observing. Many become more easily agitated and argumentative. Some say that such behavior changes can be attributed to the effects of working at high altitudes, but I think it is more likely to be due to working under pressure. There is just too much at stake.

As the night moves on and the stars move across the sky from east to west, we are constantly chasing stars that are about to set in order to get our last

(continued)

exposure on the object. This is really not the best strategy because in order to observe a setting star the telescope has to look through a lot of the Earth's atmosphere and such conditions are not optimal. The key is to stay calm and rational and be ready to let go of setting stars and just move on to a new target that has risen and is now high in the sky. This is also true in life. We have to emotionally let go of things that are gone, and welcome the new and emerging opportunities.

One of my favorite activities during the night is to go out to the cat walk around the dome on the premise of "checking the sky for cirrus and clouds." In fact, this is just an excuse to go and admire the glory of the stars in the night. Although not as spectacular as in the Southern Hemisphere, the Milky Way is prominent in the Hawaiian sky. After a few minutes to allow my eyes to adapt to the darkness (and this is total darkness as there is no artificial light anywhere in sight), I can make out faint stars that are hard to see among the light pollution of our civilized world. One has the feeling that we are as close to God and nature as one can be.

Although observing is very hard work (believe me, doing anything at 4,200 m is hard work), I often feel like a pilgrim who is privileged to be there. A pilgrimage to the stars through the telescope brings me much closer to them as I investigate their composition, conditions, surroundings, structures, and then by theoretical extrapolations, their origins and future. This feeling of empowerment is difficult to describe but very much real.

When the dome finally has to close because of the increasing sky brightness of dawn, I walk out of the building to get a glimpse of the view. The shadow of Mauna Kea is projected onto the distant sky by the sun at our back. In spite of my tiredness, this magnificent sight greatly boosts my spirits.

I first went to Mauna Kea in 1984. Over the last 20 years, my fascination with the place has not lessened. Although the sunshine and the beaches of Hawaii seem appealing, Mauna Kea draws me like a magnet that will not let go. There is something about the place that sets it apart from the rest of the world. I don't know what it is but I am certain that it is spiritual. It will be a sad day when I am too old or too frail to go through a night of astronomical observations on Mauna Kea. Until then, it is still my Mecca and Jerusalem.

Appendix A: Scientific Notation

Since in astronomy we often have to deal with large numbers, writing a lot of zeros is not only cumbersome, but also inefficient and difficult to count. Scientists use the system of scientific notation, where the number of zeros is short handed to a superscript. For example, 10 has one zero and is written as 10^1 in scientific notation. Similarly, 100 is 10^2, 100 is 10^3. So we have: 10^3 equals a thousand, 10^6 equals a million, 10^9 is called a billion (U.S. usage), and 10^{12} a trillion. Now the U.S. federal government budget is in the trillions of dollars, ordinary people really cannot grasp the magnitude of the number.

In the metric system, the prefix kilo- stands for 1,000, e.g., a kilogram. For a million, the prefix mega- is used, e.g. megaton (1,000,000 or 10^6 ton). A billion hertz (a unit of frequency) is gigahertz, although I have not heard of the use of a giga-meter. More rarely still is the use of tera (10^{12}).

For small numbers, the practice is similar. 0.1 is 10^{-1}, 0.01 is 10^{-2}, and 0.001 is 10^{-3}. The prefix of milli- refers to 10^{-3}, e.g. as in millimeter, whereas a micro-second is $10^{-6} = 0.000001$ s. It is now trendy to talk about nano-technology, which refers to solid-state device with sizes on the scale of 10^{-9} m, or about 10 times the size of an atom.

With this kind of shorthand convenience, one can really go overboard. Physicists now boldly talk about what is happening in the first 10^{-26} s of the Universe. Now that is a lot faster than a blink of an eye.

S. Kwok, *Stardust*, Astronomers' Universe, DOI 10.1007/978-3-642-32802-2,
© Springer-Verlag Berlin Heidelberg 2013

Appendix B: Units of Measurement

Distance

Because the scales in space and time that we study in astronomy are so different from our everyday lives, astronomers find it necessary to devise new, and sometimes bizarre, units of measurements. Instead of kilometer (or miles in the United States), astronomers talk about light years (the distance traveled by light in one year), or even the obscured unit of a parsec (the distance of a star which will shift in position in the sky by 1 arc second in angle when the Earth moves from one side of the Sun to the other). The unit of parsec was first devised as a common unit to refer to distance of stars, as the nearest star Proxima Centauri is located at 1.3 pc (or 4.2 light years). The distance between the Sun and the center of the Milky Way Galaxy is about 8,000 parsecs, or 26,000 light years. The Andromeda Galaxy, the nearest galaxy similar to our Milky Way galaxy, is at 770,000 parsecs (or 2.5 million light years) away. For studies of the Solar System, and now increasingly the study of extra-solar planetary systems, the astronomical unit (A.U.) is commonly used. One A.U. is the distance between the Earth and the Sun, or about 150,000,000 km. There are 63,240 A.U.s in a light year, so the size of planetary systems is much smaller than the separation between stars.

The unit light year is commonly used in popular astronomy writings, but almost never in professional literature, where the unit of parsec is almost exclusively used. Since a light year is only three times smaller than a parsec, there is no reason why light year cannot be adopted as standard unit of measurement for distance in astronomy. The practice of using parsec is more a habit and tradition than logic, as the unit light year is clearly much easier to understand. I sometimes cannot help but feel that the use of jargons in science (this is worse in social sciences) is more intended to confuse the outsiders than for clear communication.

While the size of stars are huge and the size of the Universe is even larger, in astronomy we also have to deal with small objects, such as atoms, molecules, and

S. Kwok, *Stardust*, Astronomers' Universe, DOI 10.1007/978-3-642-32802-2,
© Springer-Verlag Berlin Heidelberg 2013

the dust particles that we describe in this book. Here, more conventional systems of units are used. The size of an atom is about 0.1 nm (1 nm $= 10^{-9}$ m), and stardust that we talk about in this book has sizes of the order of a micrometer (micrometer, or μm, where 1 μm $= 10^{-6}$ m). The wavelengths of visible colors are measured in fraction of a micrometer, or hundreds of nanometer (nm).

Time

Fortunately, astronomers do not see fit to invent a new unit of time. The most commonly used units of time are the familiar units of seconds and years. However, the long lives of stars and the age of the Universe necessitate the need of using million (10^6) or billion (10^9) years to refer to long intervals of time.

Angle

The sky above us is two dimensional. We see a distribution of stars in the sky but we have no concept of depth. It is not easy to tell which star is farther away than the other. We are, however, able to measure the angular separation between two stars. Because the year is 365 days, which is close to the nice number of 360 which can be divided by 2, 3, 4, 5, 6, 8, 9, 10, 12, 15, etc., we have adopted 360° as a full circle. Again, since 60 is a good number, we divide a degree into 60 arc minute, and an arc minute into 60 arc second. A 1-cm coin placed at a distance of 1 km will have an angular size of 2 arc second, so 1 arc second is very small separation indeed.

Color

The visible light is made up of different colors ranging from red to violet. We now know that light is just part of the general phenomenon of electromagnetic waves, and there is no fundamental difference between visible light and radio waves. The only difference between the two is the wavelength (the distance between the repeating peaks of the waves), with radio waves having wavelengths of the order of meters and visible light with wavelengths of fractions of a micrometer. Within the range of visible light, different colors are characterized again by different wavelengths. The rainbow colors from red to violet correspond to a decreasing wavelength scale, with the red color having a wavelength of about 0.7 μm and the blue color about 0.4 μm. Just outside the visible range, on the long wavelengths side is the infrared, and on the short wavelength side is ultraviolet. X-rays have even shorter wavelengths than ultraviolet, with wavelengths of the order of nanometer.

Alternatively, colors can be specified by frequency measured in units of cycles per second (Hz). Frequency and wavelengths are two ways to describe the same thing, but in a reciprocal way. The multiplication product of frequency (ν) and wavelengths (λ) is a constant (the speed of light) so one can easily derive one from the other. Mathematically, the relationship between the two is

$$\nu \cdot \lambda = c,$$

where c is the speed of light (3×10^{10} cm/s). In other words, electromagnetic waves with a high frequency have short wavelengths.

Frequency as a unit of color is more commonly used in radio waves, e.g., we refer to the FM radios having frequency coverage from 88 to 108 MHz ($1\,\text{MHz} = 10^6\,\text{Hz}$). Usual cellular phone communication frequencies are 900 MHz and 1.8 GHz ($1\,\text{GHz} = 10^9\,\text{Hz}$). Expressed in wavelengths, FM radio waves have wavelengths of the order of 3 m, whereas radio waves from cellular phones have wavelengths of 33 and 17 cm. Scientists rarely ever use frequency as a measure of color in the visible.

Appendix C: Color and Temperature

The precise formula governing the amount of radiation at any wavelength emitted by an object of a certain temperature was worked out by Max Planck (1858–1947) based on the quantum nature of light. The relationship between the peak of the radiation and the temperature is very simple:

$$\lambda_{max}(cm) \cdot T(K) = 0.3$$

which is known as the Wien's law. For example, a star of 6,000° K will have its radiation peak at 0.00005 cm, or 0.5 μm, which is in the visible region. An object at room temperature (300 K) will radiate at 0.001 cm, or 10 μm, which in the infrared.

S. Kwok, *Stardust*, Astronomers' Universe, DOI 10.1007/978-3-642-32802-2,
© Springer-Verlag Berlin Heidelberg 2013

Appendix D: Naming Convention of Astronomical Objects

Since there are billions of celestial objects in the sky, there is a need to devise a system for us to refer to them. The naming of celestial objects is diverse, although the International Astronomical Union has tried to regulate the practice in recent years. The Sun, the Moon, and the Earth all have names in each individual culture. Due to historical reasons, different classes of objects are named in a different manner.

1. Planets: names of the five brightest planets Mercury, Venus, Mars, Jupiter, and Saturn were named after mystic Greek or Roman gods. The later discovered planets Uranus and Neptune also followed this tradition. Outside of the western culture, all these planets have different names in different languages.
2. Asteroids: the early asteroids were thought to be minor planets and therefore were given names in a similar way as the planets. Examples are Ceres, Pallas, Juno, Vesta, etc. The discoverer is given the opportunity to propose to the International Astronomical Union (IAU) to give a name for the discovered object. With modern automatic telescopes, thousands of asteroids can be discovered over very short time periods and the naming of asteroids has lost its meaning. With hundreds of thousands of asteroids known and the number increasing rapidly, most of them will never be named.
3. Satellites of planets: the satellites of Mars, Phobos and Deimos were named after sons of the gods of Mars. The four moons of Jupiter found by Galileo (Io, Europa, Ganyemde, and Callisto) were named after lovers of Jupiter (a Roman mythical god, or Zeus in Greek). Current convention is that newly discovered moons of Jupiter will follow this tradition. The early discovered satellites of Saturn (e.g., Enceladus, Titan) are named after Titans or their descendants. The more recently discovered satellites of Saturn are named after giants and monsters in other mythologies. The satellites of Uranus are drawn from characters in the English literature, in particular plays of William Shakespeare. Some examples are Titania and Oberon from "*A Midsummer Night's Dream*", Miranda from "*The Tempest*". With increasing level of technology, smaller and smaller satellites can be discovered and it may become impractical to name them all.

S. Kwok, *Stardust*, Astronomers' Universe, DOI 10.1007/978-3-642-32802-2,
© Springer-Verlag Berlin Heidelberg 2013

4. Comets: comets are generally named after their discoverers (e.g., comet Hale-Bopp).
5. Stars: only some of the bright stars have recognized names. Examples include Vega, Sirius, Canopus. Other star names are derived from catalogues. For example the Bayer catalogues have 1,564 bright stars, using a combination of Greek letters and constellation names (e.g., α Ori, which is Betelgeuse and α Lyr is Vega). Other star catalogue names use a combination of letters and numbers, and frequently employ the coordinates of the stars as part of the name. Some commonly used catalogues include the Bonner Durchmusterung (BD), the Henry Draper Catalog (HD), the bright star catalog (HR), so a star can therefore have many names. Betelgeuse is α Ori, and is also HR 2061, BD + 7 1055, HD 39801.

   Variable stars are named by their constellations with preceding letters given in order of discovery date and arranged in order similar to car license plates. For example, CW Leo is a variable star in the constellation of Leo and got the initial CW based on the time it got the variable star designation.
6. Nebulae, galaxies, and star clusters: these extended objects usually are labeled by their catalogue names, e.g., NGC 7027 is number 7027 in the New General Catalogue of Nebulae and Clusters of Stars compiled by John Louis Dreyer in 1888. The NGC catalogue has 7,840 objects, which was later supplemented by 5,386 objects in the Index Catalogues (IC) in 1895 and 1908.

Appendix E: Elemental Abundance

All chemical elements are either made in the Big Bang or in stars. However, the distribution of the elements among different bodies is uneven. The Solar System, having condensed from an interstellar cloud, has approximately the same elemental abundance as the Milky Way galaxy. However, the Earth (as in other terrestrial planets) cannot keep the lighter gases and its solid body has a very different chemical make up from the Sun. The human body, being a special chemical machine, is also different. In this table, we show the distribution of some common elements by mass fraction in the Galaxy, the Solar System, Earth, and the human body.

Sometimes the abundance fractions are expressed in numbers of atoms. To convert from mass abundance to number abundance, divide the mass abundance by the atomic weight of each of the elements and take the fraction of the total. For example, the number abundance of H, He, and O are 90, 9, and 0.08 %, respectively.

Element	Atomic number	Atomic weight	Cosmic abundance	Solar system	Earth	Human body
			Mass fraction (%)			
Hydrogen (H)	1	1	73.9	70.6	0.03	10
Helium (He)	2	4	24.0	27.5	–	–
Oxygen (O)	8	16	10.4	5.92	29.7	65
Carbon (C)	6	12	4.60	3.03	0.07	18
Neon (Ne)	10	20	1.34	1.55	–	–
Iron (Fe)	26	56	1.09	1.17	31.9	<0.05
Nitrogen (N)	7	14	0.96	1.11	0.003	3
Silicon (Si)	14	28	0.65	0.65	16.1	–
Magnesium (Mg)	12	24	0.58	0.51	15.4	0.05
Sulfur (S)	16	32	0.44	0.40	0.64	0.2

S. Kwok, *Stardust*, Astronomers' Universe, DOI 10.1007/978-3-642-32802-2,
© Springer-Verlag Berlin Heidelberg 2013

Appendix F: Mass and Energy

The amount of energy release in an external impact can be estimated from simple physics. Assuming a typical incoming asteroid has a density (ρ) of rock, or about 3,000 kg per cubic m. If the size (r) of the asteroid is 50 m, then the mass of the asteroid (M) is

$$M = \frac{4}{3}\pi\rho r^3$$

or about 1.6 million metric tons.

For an object to leave the gravitational attraction of the Earth, it must have the kinetic energy needed to overcome the gravitational potential at the surface of the Earth. The minimum velocity, called the escape velocity, is given by

$$v_{escape} = \sqrt{\frac{GM_{earth}}{R_{earth}}}$$

where G is the gravitational constant (6.67300×10^{-11} m^3/kg s^2), M_{earth} is the mass of the Earth (5.97×10^{24} kg), and R_{earth} is the radius of the earth (6,378 km). The escape velocity of the Earth is therefore 11.2 km/s. Since the incoming asteroid is falling into the Earth, when it strikes the Earth, it will have a speed at least equal to the escape velocity. Most of the time, the asteroid will be coming with some initial velocity, so let us assume that the final speed (v) of the asteroid upon collision with the Earth's surface is 15 km/s. The final kinetic energy of the asteroid is

$$KE = \frac{1}{2}Mv^2$$

which is about 1.8×10^{17} J. Since one ton of TNT releases about 4.2×10^9 J, the amount of energy release by the asteroid impact is about 40 million tonnes of TNT. In comparison, the atomic bomb that was dropped on Hiroshima had a yield of

S. Kwok, *Stardust*, Astronomers' Universe, DOI 10.1007/978-3-642-32802-2,
© Springer-Verlag Berlin Heidelberg 2013

about 15 kilo tonnes of TNT and a typical hydrogen bomb has a yield of several megatonnes of TNT.

The actual energy release is dependent on the angle of the impactor hits the Earth, and also on whether it hits water, ice, porous rock, or dense rock. But one can get an idea of the effect of impact based on the above simple calculations.

A similar calculation can be done for cometary impact. Since comet is mostly ice, the density of a comet is lower, or about 1,000 kg/m^3. If the comet has a radius of 10 km, then the mass of the comet is about 4 trillion metric tons. Comets have their own orbital velocities around the Sun and therefore usually come in faster. Let's assume that the incoming velocity is 50 km/s. The amount of energy release is 10^{15} tonnes of TNT (or a billion megaton of TNT), which is a staggering number. The effects of such a cometary impact will be devastating.

Glossary

Scientists develop technical terms in order to be more precise in the meaning. These terms ("jargons") represent a major obstacle for laymen to understand scientific literature. And there are the acronyms. The cynical will say that they are designed to confuse outsiders. Another unfortunate fact is that some technical terms are also words used in common language (e.g., force, energy, theory), but they have very different meanings in science. In recent years, there are also cases where scientists deliberately use common words to describe a totally different physical phenomenon, which causes further confusion. Some examples are "black holes", which really means a gravitational singularity and is neither black nor is a hole; and "dark energy", which is really negative pressure. Astronomers use the term "burning", which in our everyday usage represents a chemical reaction involving oxygen, to stand for nuclear reactions in stars. "Metals" to an astronomer is not a shiny, hard, substance like iron or copper, but any element heavier than helium. In this book, I have tried to minimize the use of technical terms, but this cannot be totally avoided. This glossary is put together for easy reference when readers encounter an unfamiliar word.

Air A mixture of oxygen and nitrogen molecules in the Earth's atmosphere. Air is essential for human life as we need oxygen molecules for respiration.

Air Force Geophysics Laboratory (AFGL) The Air Force Cambridge Research Laboratory (CRL) in Hanscom Field was created in 1949. In 1974, it was renamed the Air Force Geophysics Laboratory. The U.S. Air Force took a great interest in infrared astronomy, in particular in the survey of the infrared sky. Research AFGL supported the AFGL sky rocket survey and the *Midcourse Space Experiment (MSX)* space mission which performed an infrared survey of the Galactic plane.

Aliphatics The term "aliphatics" is derived from the Greek word "aleiphas", meaning "fat". Common animal fats and oils are made of long chains of hydrocarbons.

Aromatics The class of aromatic molecules has its origin in chemical extracts from pleasant smelling plants. These are hydrocarbons with a ring-like structure.

S. Kwok, *Stardust*, Astronomers' Universe, DOI 10.1007/978-3-642-32802-2,
© Springer-Verlag Berlin Heidelberg 2013

Atmosphere A layer of gas that covers the surface of a planet. Although there is no definite outer boundary of the atmosphere, the altitude of 100 km is commonly used as the line separating the atmosphere and outer space. The Earth's atmosphere is made up of primarily nitrogen (78 %) and oxygen (21 %) molecules. However, the chemical composition of the Earth's atmosphere has greatly changed from the early days of the Earth. The emergence of oxygen is the result of life.

Atmospheric Pressure Because of gravity, the air in the Earth's atmosphere also has weight and we feel this weight as atmospheric pressure. As we climb to higher altitudes, there is less amount of air above us, and the pressure we feel is less. As we go under water, we suffer from the additional pressure from the water. Under the ocean at a depth of 3 km, the pressure can amount to 300 times the atmospheric pressure at sea level.

Astrobiology The study of possibility of life beyond the terrestrial environment. The field encompasses several sub disciplines, including the search for life in the Solar System, search for signals from extraterrestrial intelligence by remote sensing, and the study of the origin of life.

Astrochemistry The discipline of studying chemical processes in space.

Astromineralogy An interdisciplinary science between astronomy and geology on the study of minerals made by stars.

Astronomical Unit (AU) The distance between the Earth and the Sun.

Asymptotic Giant Branch (AGB) Also called the "2nd giant branch" is a phase of stellar evolution after the red giant phase ("first giant branch"). AGB stars are very large (hundreds of solar radii) and luminous (thousands of solar luminosity) and have surface temperatures of only a few thousand degrees Kelvin. The internal structure of an AGB star consists of a dense hot core made of carbon and oxygen, and a very extended envelope made of hydrogen. The AGB is the phase of evolution where most of the heavy chemical elements are synthesized by nuclear reactions. AGB stars are often pulsating and generally have a stellar wind.

Bacteria Single cell microorganisms without a membrane-bound nucleus. Bacteria are found everywhere on Earth, including our digestive tracts and in deep-sea vents.

Bioluminescence The production of light from a chemical reaction occurring within a living organism.

Biomolecules Biomolecules are molecules in living organisms that are directly responsible for the biological functions. Biomolecules and reactions among them for the basis of the discipline of biochemistry.

Biosphere Parts of the Earth where life exists. These include the surface of the Earth, the oceans, and the lower atmosphere.

Burning refers to chemical reactions involving a fuel and oxygen, producing heat as a result. The technical term is combustion. Unfortunately, astronomers use the word "burning" to refer to nuclear fusion reactions, e.g., "hydrogen burning", or "helium burning", in stars. These are nuclear reactions and are not chemical reactions.

Coal Remnants of ancient life. After under heat and pressure for millions of years underground, it developed into a complicated organic solid now used primarily for fuel.

Carbon Star Evolved stars whose surface abundance of carbon exceeds that of oxygen. This occurs at the very late stages of stellar evolution as the element carbon is synthesized in the interior and brought up to the surface by convection.

Chlorophyll Derived from the Greek word for "green leaves", chlorophyll is the molecule in plants that is responsible for photosynthesis. Since it absorbs most of the blue and red color sunlight, it is seen as green color from the reflected light.

Cirrus Clouds Cirrus refers to the light, diffuse clouds in the atmosphere. This term is also used to refer to diffuse interstellar clouds in the Galaxy.

Comets Pieces of rock and ice left over in the outer Solar System after its formation. They sometimes venture into the inner Solar System and when heated by sunlight, develop spectacular appearing tails.

Chondrites Meteorites that have not been modified by melting. Of particular interest are the carbonaceous chondrites, which represent about 5 % of all chondrites.

Crater The term impact crater refers to the mark left from an impact of an external body colliding with the Earth. Although similar in shape, their origins are quite different from volcanic craters which are created as the result of volcanic eruptions.

Diamond A mineral made up of carbon atoms arranged in a tetrahedral structure.

Dispersion When light passes through a medium, different colors are affected differently. This color dependence effect is called dispersion. The most common example of dispersion is the rainbow, when the white sunlight is dispersed by water vapor in the atmosphere.

Doppler Effect The change of light color as the result of movement of the emitting object. An object moving towards or away from us will have its light shift to blue or red respectively.

Dust In astronomical literature, the term "dust" is used to refer to micrometer-size, inorganic or organic solid-state particles. The "dust" in our daily environment contains a large fraction of biological materials, which are not assumed to be present in interstellar or stellar "dust".

Ecliptic The annual path of the Sun in the sky across the stellar constellations. The ecliptic is also the path taken by the planets. The 12 constellations on the ecliptic (called the Zodiac) were believed by the ancient people to have particular significance as they are favored by the Sun.

Energy A precise term used by physicists to measure the capacity for an object to do work. Energy can be in different forms (e.g., gravitational, chemical, thermal, or nuclear) and can be converted from one form to another. For example, the contraction of a star causing the star to heat up (gravitation to thermal), the burning of coal can generate heat (chemical to thermal). The breakdown of the hydrocarbon molecules in gasoline allows the chemical energy stored in the molecules to be used to drive a vehicle into motion (kinetic energy).

Enzymes Substances that promote chemical reactions in biological systems. Enzymes are needed, e.g., in the digestion of food, where proteins, carbohydrates and fats are broken down into smaller molecules.

Eukaryotes Organisms with cells that contain a nucleus.

Exogenous Delivery The delivery of materials to Earth from the outside.

Extremophiles Organisms that prefer to live in very high or very low temperature, high pressure, or high salt content surroundings.

Fossils Marks left by dead animals and plants on rocks. Since flesh decomposes with time but rocks can last millions or even billion of years, our learning of life forms in the past rely on such imprints.

Fullerene A molecule made up of 60 carbon atoms in the shape of a soccer ball. It was first artificially created in the laboratory in 1985 and is considered as the third form of carbon, after graphite and diamond, two forms of pure carbon that exist naturally.

Frequency A measure of the energy of light. Frequencies are measured in units of Hertz (cycles per second).

Fusion The nuclear process that combines light elements into heavier ones. Fusion is the principle behind the hydrogen bomb and is the source of power of stars.

Gaia Hypothesis A theory proposed by James Lovelock that treats the biological and non-biological parts of the Earth as one integrated system. The life, air, rocks and water components of the Earth are coupled and related to each other, and are all evolving as one system.

Gem (or Gemstone) A piece of mineral used to make jewelry. Other than their color and reflective properties, rarity is another factor that determines the value of gems. Some organic materials such as amber can also be considered as gems.

Heavy Bombardments The early Solar System was much more chaotic than now. Fragments left over from the formation of the Solar System often collided with the newly formed planets, and the Earth had suffered from such heavy bombardments. The heavy bombardment period ended about 3.8 billion years ago.

Heavy Elements This term (also the term "metals") are used in astronomy to refer to chemical elements heavier than helium.

Hydrocarbons The family of organic compounds that contain only hydrogen and carbon. Some examples of simple hydrocarbon are methane (CH_4), ethane (C_2H_6), and propane (C_3H_8). More complicated forms of hydrocarbons include animal fats, which contain long chains of hydrocarbons.

Hydrolysis By adding water to a substance, some of the chemical bonds of the substance can be broken, allowing molecular fragments to be separated.

Hydrothermal Vents High temperature water flows in the deep sea floor. They attract strong interests because of the biological colonies found in them. This demonstrates that life can exist independent of energy from the Sun.

Ice A solid, crystalline form of water. Other molecules, such as methane and carbon dioxide can also have an ice form at low temperatures.

Igneous Rock (from Latin Ignis Meaning Fire) is formed by solidification of molten rock. Examples: granite, basalt.

Interplanetary Dust Particles (IDP) Micrometer-size particles originated from asteroids or comets.

Insoluble Organic Matter (IOM) The insoluble component of organic matter in carbonaceous meteorites.

Interstellar Medium The general space between stars. The Milky Way galaxy contains about 100 billion stars, but the stars are widely separated from each other. The space in between, however, is not totally empty. There are low concentrations of gas and dust in the vast volumes between stars. These materials are referred to as the interstellar medium.

Kuiper Belt A disc-shape region in the outer Solar System from the orbit of Neptune to about 55 A.U. from the Sun. Over a thousand objects in the Kuiper Belt have been found, of which Pluto is the most well-known member.

Lipids Lipids are water insoluble hydrocarbons. The most familiar examples are animal fats and vegetable oil.

Lithosphere The solid crust of the Earth. This includes the continents and the floor of the oceans.

Low Resolution Spectrometer (LRS) A Dutch-built instrument onboard the *IRAS* satellite.

Main Sequence The period of a star's life that relies on nuclear fusion of hydrogen into helium in the core for power.

Mantle of the Earth A rocky shell of about 2,900 km thick which separates the core and the crust of the Earth.

Metamorphic Rocks (from Greek Metamorphosis Meaning Transform) are the result of rocks being transformed by heat or pressure. Marble is one example of metamorphic rocks.

Meteors Meteors refer to the optical phenomenon of flashes of light in the sky due to solid particles passing through the atmosphere and being vaporized in the process.

Meteoroids Meteoroids are the general term that refers to solid particles in interplanetary space, with sizes between 100 and 10 m.

Meteorites Meteorites are remnants of solid objects that pass through the atmosphere, strike the surface of the Earth, and survive the impact. The relationships between these three terms are as follow: a meteoroid enters the atmosphere, being seen as a meteor, strikes the ground and leaves behind a meteorite.

Micronmeter (or micron, μm) One millionth (10^{-6}) of a meter.

Micrometeorites Small meteorites

Microwave Microwave refers to electromagnetic radiation between infrared and radio radiation. There are no precise boundaries between infrared/microwave, or microwave/radio. But generally microwave refers to radiation with wavelengths roughly between 1 m and 1 mm.

Millimeter Wave Astronomers use the term "millimeter wave" to refer to electromagnetic radiation with wavelengths of the order of 1 mm. This wavelength region is where many of the rotational transitions of molecules lie. The lowest rotational transition of the CO molecule is at the wavelength of 2.6 mm. The term "sub-millimeter-wave" refers to radiation with wavelengths shorter than 1 mm. The most advanced telescope operating in these wavelength regions is the Atacama Large Millimeter/submillimeter Array (ALMA).

Mineral A naturally occurring element or chemical compound, usually an inorganic substance with crystalline structure.

Mira Variables A class of variable stars named after the prototype star Omicron Ceti in the constellation of Cetus. The varying brightness of these stars is the result of the pulsation of the stellar atmosphere. These are very old, evolved stars with red colors and luminosities thousands of times that of the Sun.

Molecules Stable entities formed by two or more atoms. The study of molecules forms the basis of the discipline of chemistry.

Moon The only natural satellite of the Earth and the second most luminous object in the sky after the Sun. The visible brightness of the Moon is entirely due to reflected sunlight, but the Moon does shine on its own in the infrared. The Moon is believed to be created after a collision between a Mars-sized object with the Earth. The materials torn off from the surface of the Earth accumulated near the Earth orbit to form the Moon.

NASA National Astronautic and Space Administration. This is the space agency of the United States of America that supports many of the science exploration missions.

Nanodiamonds Diamonds of nanometer size. Most are artificially synthesized by natural nanodiamonds can also be found in impact sites and in meteorites.

Nanometer (nm) One billionth (10^{-9}) of a meter.

Nanoparticles Solid-state dust grains of sizes of the order of nanometers. The term is not precisely defined and has been applied loosely to particles with diameters between 1 and 100 nm.

Near Earth Objects (NEO) Meteoroids that cross the Earth's orbit and may have a chance of hitting the Earth.

Nebulae The Latin word for "clouds" or "mist". In astronomy, it is used for objects that have an extended angular size in the sky, unlike stars or planets which seem to be point-like. The original term "nebulae" included both gaseous nebulae and collection of stars. The latter group now uses the terms "star clusters" or "galaxies" and nebulae are now reserved to gaseous nebulae, such as "planetary nebulae" and "reflection nebulae".

Novae A form of exploding stars that brightens on time scale of days. Novae are the result of binary evolution where matter is accreted to the surface of a white dwarf from its companion. When sufficient materials accumulate, nuclear reaction converting hydrogen into helium is initiated, resulting in a sudden increase in the stellar luminosity.

Oort Cloud A spherical region of radius up to 50,000 A.U. from the Sun. The Oort Cloud is believed to be a reservoir of comets.

Panspermia has its origin in the Greek word which can be translated to mean "seeds everywhere". Now it is used to refer to the hypothesis that life can be transported across the Universe and be established under suitable conditions.

Paraffin The term paraffin is derived from the Latin parum affinis (slight affinity) as paraffins are generally inert to chemical reagents. In chemistry, the term "paraffin" is equivalent to "alkanes", which is a family of saturated hydrocarbons. Some familiar examples are methane (CH_4) and octane (C_8H_{18}).

Peptide A very short protein, made of a few amino acids

Photometry The astronomical technique to measure the brightness of small celestial objects. Since stars and distant galaxies have small angular sizes, all their light is essentially concentrated in one point. The measurement of the brightness of small light sources (called "flux" in technical terms) is different from our usual concept of brightness in everyday life, which is distributed over an extended area (called "intensity" in technical terms).

Planetesimals Small sized (1–10 km) bodies formed out of the solar nebula. Planets are believed to have formed from aggregation as the result of collisions between planetesimals.

Planetary Nebulae Gaseous nebulae ejected and photoionized by a star evolving between the AGB and white dwarf phases of stellar evolution.

Plasma Refers to the ionized state of matter when one or more electrons are detached from the atom. Stars are mostly in a plasma state.

Polycyclic Aromatic Hydrocarbons (PAH) Molecules made up of several joined rings of only carbon and hydrogen.

Polymers Chemical compounds containing multiple units of similar chemical structures.

Prebiotic Molecules A subset of organic molecules which are the structural basis of biomolecules, which are molecules directly responsible for the functions of life. Prebiotic molecules can be considered as precursors to biomolecules in the pathway to the origin of life.

Primordial Soup The liquid environment consisting of organic molecules from which life arose on the primeval Earth.

Prokaryotes Organisms with cells that are without a nucleus. Most of the prokaryotes have only one cell in the organism.

Proteins Long polymers of amino acids.

Proto-Planetary Nebulae The immediate precursor of planetary nebulae. Proto-planetary nebulae are distinguished from planetary nebulae in that the nebulae are not ionized and they shine strictly by reflection of light from the central star.

Red Giants The evolutionary stage of stars after the Main Sequence. When hydrogen in a stellar core is exhausted by nuclear burning, a star begins to burn hydrogen in an envelope surrounding the helium core. The stellar envelope expands and the star becomes more luminous.

Reflection The change in direction of light upon hitting a surface of matter. Our visual perception of objects are mostly based on reflected light.

Refraction The change in direction of light as the result of light passing from one medium to another. The most common examples of refraction are light entering water or passing through a piece of glass.

Rotation A molecule can rotate around a common axis. The simplest example one can have is a two-atom molecule such as CO. The C and O atoms rotate around an axis perpendicular to the line connecting the two atoms. Quantum principles dictate that the molecule can have only discrete rotational speeds. When they change from a higher to lower rotation rate, radio waves (usually in the millimeter wavelength range) are emitted.

Satellites Natural bodies that revolve around planets. For example, Titan is a satellite of Saturn. It is also referred to as a moon of Saturn as the Moon is a satellite of the Earth. Artificial satellites are man-made objects that are sent into orbits around the Earth. An artificial satellite can in principle stay in orbit forever, but low-orbit satellites suffer from atmospheric drag and will eventually fall back on Earth.

Spatial (or Angular) Resolution The ability of telescopes to see the fine details in an astronomical image. We are familiar with this concept when we use digital cameras. The more pixels in the CCD of a digital camera, the more details one can capture in a picture. For astronomical observations, higher spatial resolution can be achieved by a larger telescope. However, telescopes on the surface of the Earth suffer from the natural limitation of imposed by the atmosphere. For this reason, space-based telescopes such as the *Hubble Space Telescope* can take sharper pictures.

Sedimentary Rock The term is derived from Latin word sedimentum, meaning settling. They are formed by accumulation of debris through breakdown of older rocks by wind and water. Sandstone and limestone are examples of sedimentary rock.

Silicate Silicate is a mineral that is commonly found on the surface of the Earth. Silicate is primarily made up of elements silicon and oxygen, but can contain other metals such as iron, magnesium, and so on. An amorphous form of silicates is commonly found in old, oxygen-rich, red giant stars.

Smoke Solid particles generated as the result of burning. Examples include the burning of wood, candles, oil lamps, etc. When the burning is not complete, the surplus carbon atoms congregate themselves into particles that make up smoke.

Soot Carbon particles produced by incomplete combustion of hydrocarbons. The most commonly observed example is automobile exhaust generated by internal combustion engine (in particular diesel engines).

Spectroscopy A technique of splitting light into small intervals of color. Since atoms and molecules emit light of unique frequencies, spectroscopy is the technique that allows us to detect atoms and molecules in space.

Stardust Solid particles made by stars. Stardust can be mineral like (e.g., silicates), or carbonaceous (organic) in composition. Their sizes range from tens of nanometer to a micron.

Steam Gaseous form of water. Water is transformed to steam (boiling) at the temperature of $100°$ C under atmospheric pressure.

Stellar Atmospheres Since stars are gaseous objects, the term "atmosphere" has a different meaning from the atmosphere of the Earth. The Earth has a solid surface, and the gaseous atmosphere is easily distinguishable. In a star, there is no clear physical separation between the atmosphere and the star itself. Instead, astronomers use the concept of "stellar atmosphere" to refer to the top layer of a star which we can see through. Beyond a certain depth, the atmosphere is no longer transparent. This point is called the "photosphere" which refers to the surface of the star. It is important to realize that the photosphere is not a physical surface.

Stellar Winds Streams of matter (ions, atoms, molecules, solids) flowing from the surface of stars. Although most stars (including the Sun) have stellar winds, the magnitudes of winds are strongest in old red giants and young, massive stars.

Surveys An unbiased observation (imaging or spectroscopic) of a wide area of the sky. This is in contrast to the common mode of astronomical observations which is targeted at specific objects.

Tar A black liquid of high viscosity derived from coal.

TNT Megaton of TNT is used as a measure of energy release during an explosion. Since the release of chemical energy of TNT varies, one megaton of TNT is artificially defined as 4.184 giga joules

Ton In the UK, one ton is 2,240 lb and in the U.S. and Canada, one ton is 2,000 lb (907 kg). These are different from the metric ton (or tonnes), which is 1,000 kg.

Unidentified Infrared Emission Features A set of features in the infrared spectra of nebulae and galaxies that arise from emission by organic compounds.

Ultraviolet Light beyond the color blue that can be seen by the human eye.

Vibration A molecule can undergo vibrational motion such as stretching and bending. For a two-atom molecule, the link between the two atoms can stretch into different lengths. A three or more atom molecule can also bend, meaning that the angles between atoms can change. A change from a higher rate of stretch (or bend) to a lower rate of stretch (or bend) gives off infrared light of characteristic wavelengths, therefore allowing us to identify the motion.

Wavelength A quantitative measure of color. The human eye response range extends from red (corresponding to a wavelength of approximately 0.7 μm) to violet (a wavelength of approximately 0.3 μm).

Zodiacal Light A band of light in the plane of the ecliptic seen usually in the west after twilight or in the east before dawn. Zodiacal light is the result of reflection of sunlight by small solid particles in the ecliptic plane.

Bibliography

The following references are provided for readers who wish to obtain more details on the subjects covered in this book.

Chapter 1: Astrobiology

The book by Shklovskii and Sagan was the first serious modern scientific discussion on the subject of astrobiology, bringing together the disciplines of astronomy and biology.

Shklovskii, I.S., and Sagan, C. 1966, *Intelligent Life in the Universe*, San Francisco: Holden-Day

Chapter 1: Review of the Current Developments of Organic in Space

Kwok, S. 2011, *Organic Matter in the Universe*, Wiley, New York

Chapter 1: Laboratory Synthesis of Organic Compounds

Simoni, R.D., Hill, R.L., Vaughan, M. 2002, Urease, the first crystalline enzyme and the proof that enzymes are proteins: the work of James B. Sumner, *J. Biological chemistry*, **277**, e23

Chapter 1: Vitalism

Rosenfeld, L. 2003, William Prout: Early 19th Century Physician-Chemist, *Clinical Chemistry*, **40:4**, 699–705.

Chapter 1: The Miller-Urey Experiment

Miller, S.L. 1953, Production of amino acids under possible primitive earth conditions, *Science*, **117**, 528–529.
Miller, S.L., Urey, H.C. 1959, Organic Compound Synthesis on the Primitive Earth, *Science*, **130**, 245–251.
Shapiro, R. 1986, *Origins; A Skeptics Guide to the Creation of Life on earth*, Summit Books, New York.

Chapter 1: Theory of Panspermia

Hoyle, F., and Wickramasinghe, C. 1984, *From Grains to Bacteria*, University College Cardiff Press, Bristol
Hoyle, F., and Wickramasinghe 1999, Comets—A Vehicle for Panspermia, *Astrophysics and Space Science*, **268**, 333

Chapter 2: The Solar System

McFadden, L.-A., Weissman, P.R., and Johnson, T.V. 2007, *Encyclopedia of the Solar System* (2nd edition), Elsevier
Davis, A.M., Holland, H.D. 2004, *Treatise on Geochemistry, Vol. I: Meteorites, Comets, and Planets, Elsevier*

Chapter 2: Origin of Zodiacal Light

Nesvorný, D., Jenniskens, P., Levison, H.F., Bottke, W.F., Vokrouhlický, D., & Gounelle, M. 2010, Cometary Origin of the Zodiacal Cloud and Carbonaceous Micrometeorites. Implications for Hot Debris Disks. *Astrophysical Journal*, **713**, 816–836.

Chapter 2: Historical Meteorites

Marvin, U.B. 1995, Siena, 1794: History's Most Consequential Meteorite Fall, *Meteoritics*, **30**, 540-541.

Chapter 2: A Story on the Prediction of the Fall and Recovery of a Meteorite

Kwok, R. 2009, The rock that fell to earth, *Nature*, **458**, 401–403

Chapter 2: Discovery of KBO

Jewitt, D.C., and Luu, J.X. 1993, Discovery of the Candidate Kuiper-Belt Object, *Nature*, **362**, 730–732.
Biography of Jewitt and Luu 2012, Shaw Prize Foundation

Chapter 2: Reviews of the Subject of Meteorites

Norton, O.R. 2002, *The Cambridge Encyclopedia of Meteorites*, Cambridge University Press
Bevan, A., and de Laeter, J. 2002, *Meteorites: a journey through space and time*, Smithsonian

Chapter 2: Discovery of Organics in Meteorites

Nagy, B., Hennessy, D.J., and Meinschein, W.G. 1961, Mass Spectroscopic Analysis of the Orgueil Meteorite: Evidence for Biogenic Hydrocarbons, *Annals of the New York Academy of Sciences*, **93**, 27–35
Bernal, J.D. 1961, Significance of Carbonaceous Meteorites in Theories on the Origin of Life, *Nature*, **190**, 129–131.
Cronin, J.R., Pizzarello, S., Cruikshank, D.P. & Matthews, M.S. 1988, Organic Matter in Carbonaceous Chondrites, Planetary Satellites, Asteroids and Comets, in *Meteorites and the Early Solar System*, ed. John F. Kerridge, 819–857, Arizona
Cronin, J.R., and Chang, S. 1993, Organic Matter in Meteorites: Molecular and Isotopic and Analyses of the Murchison Meteorite, in *The Chemistry of Life's Origin*, ed. J.M. Greenberg, C.X. Mndoza-Gómez and V. Pirronello, (Dordrecht: Kluwer), pp. 209–258.

Chapter 3: Origin of Craters on the Moon

Wilhelms, D.E. 1993, *To a Rocky Moon: a Geologist's History of Lunar Exploration*, The University of Arizona Press, Arizona
Wegener, A. 1921, English translation, 1975, The Origin of Lunar Carters, *The Moon*, **14**, 211–236

Chapter 3: Number of Impact Craters on Earth

Grieve. R.A.F. 2001, Impact cratering on Earth, *Geological Survey of Canada Bulletin* 548: A Synthesis of Geological Hazards in Canada, p. 207–224.
Gomes, R., Levison, H.F., Tsiganis, K., & Morbidelli, A. 2005, Origin of the Cataclysmic Late Heavy Bombardment period of the Terrestrial Planets, *Nature*, **435**, 466–469.

Chapter 3: Discovery of the Xiuyan Crater in China

Chen, M., Koeberl, C., Xiao, W., Xie, X., & Tan, D. 2011, Plana Deformation Features in Quartz from Impact-produced Polymict Breccia of the Xiuyan Crater, China, *Meteoritics & Planetary Science*, **46**, 729–736.

Chapter 3: Exogenous Delivery

Chyba, C.F., and Sagan, C. 1992. Endogenous Production, Exogenous Delivery and Impact-shock Synthesis of Organic Molecules: an Inventory for the Origins of Life, *Nature* 355, 125–132.

Chapter 3: External Impact as the Cause of the Mass Extinction of 65 Million Years Ago

Alvarez, L.W., Alvarez, W., Asaro, F. & Michel, H.V. 1980, Extraterrestrial Cause for the Cretaceous-Tertiary Extinction, *Science,* **208**, 1095–1108.

Chapter 3: Discovery of the Chicxulub Crater

Hildebrand, A.R. et al. 1991, Chicxulub Crater: a Possible Crestaceous/Tertiary Boundary Impact Crater on the Yucatan Peninsula, Mexico. *Geology*, **19**, 867–871.

Chapter 3: The Permian/Triassic Event

Becker, L., Poreda, R.J., Hunt, A.G., Bunch, T.E. & Rampino, M. 2001, Impact Event at the Permian-Triassic Boundary: Evidence from Extraterrestrial Noble Gases in Fullerenes. *Science* **291**, 1530–1533.
Braun, T., Osawa, E., Detre, C., Tóth, I. 2001, On Some Analytical Aspects of the Determination of Fullerenes in Aamples from the Permian/Triassic Boundary Layers, *Chemical Physics Letters*, **348**, 361–362.
Becker, L. *et al.* 2004, Bedout: A Possible End-Permian Impact Crater Offshore of Northwestern Australia. *Science* **304**, 1469–1476.

Chapter 4: The Life of Fred Hoyle

Hoyle, F. 1994, *Home is where the wind blows: chapters from a cosmologist's life*, University Science Books, Philadelphia
Mitton, S. 2005, *Conflict in the Cosmos: Fred Hoyle's Life in Science*, Gardners Books, United Kingdom

Chapter 4: Theories on the Origin of Elements

Alpher, R.A., Herman, R., Gamow, G.A. 1948, Thermonuclear Reactions in the Expanding Universe, *Physical Review*, **74**, 1198–1199.
Burbidge, E.M., Burbidge, G.R., Fowler, W.A., & Hoyle, F. 1957, Synthesis of the Elements in Stars, *Reviews of Modern Physics*, **29**, 547–650.
Hoyle, F., Fowler, W.A., Burbidge, G.R. & Burbidge, E.M. 1956, Origin of the Elements in Stars. *Science* **124**, 611–614.
Cameron, A.G.W., 1957, Stellar evolution, nuclear astrophysics, and nucleogenesis, Chalk River Report CRL-41.

Chapter 4: Nuclear Processes Responsible for the Synthesis of Chemical Elements

Wallerstein, G. et al. 1997, Synthesis of the Elements in Stars: Forty Years of Progress, *Review of Modern Physics*, **69**, 995–1084.

Chapter 4: Personal Accounts of Fowler and Cameron

Fowler, W.A. 1983, *The Quest for the Origin of the Elements*, Nobel Prize Lecture
Cameron, A.G.W. 1999, Adventures in Cosmogony, *Annual Review of Astronomy & Astrophysics*, **37**, 1–36.

Chapter 5: History of Infrared Astronomy

Gehrz, R.D. *The History of Infrared Astronomy: the Minnesota-UCSD-Wyoming Axis*, unpublished manuscript
Price, S.D. 2009, Infrared Sky Surveys, *Space Science Reviews*, **142**, 233–321.

Chapter 5: History of the Discovery of Infrared Light

Chang, H., and Leonelli, S. 2005, Infrared Metaphysics: the Elusive Ontology of Radiation. Part 1, *Study in History and Philosophy of Science*, **36**, 477–508.

Chapter 6: Dust in Comets

Becklin, E.E., and Westphal, J.A. 1966, Infrared Observations of Comet 1965f, *Astrophysical Journal*, **145**, 445–453.

Chapter 6: prediction of existence of star dust

Hoyle, F. 1955, Frontiers of Astronomy,London: Heinemann

Chapter 6: Detection of Amorphous Silicates in Stars

Woolf, N.J., and Ney, E.P. 1969, Circumstellar Infrared Emission from Cool Stars, *Astrophysical Journal*, **155**, L181-184.

Chapter 6: Detection of Silicon Carbide in Stars

Treffers, R. and Cohen, M. 1974, High-Resolution Spectra of Cool Stars in the 10 and 20-micrometer Regions, *Astrophysical Journal*, **188**, 545–552.

Chapter 6: Stars Showing Silicates and Silicon Carbide from Observations by the Low Resolution Spectrometer on the *IRAS* Satellite

Kwok, S., Volk, K., & Bidelman, W.P. 1996, Classification and Identification of IRAS Sources with Low-Resolution Spectra, *Astrophysical Journal Supplements*, **112**, 557–584.

Chapter 7: The Discovery of Extreme Carbon Stars

Volk, K., Kwok, S., and Langill, P.P. 1992, Candidates for Extreme Carbon Stars, *Astrophysical Journal*, **391**, 285–294.

Chapter 8: The *ISO* Mission

Kessler, M.F., et al. 1996, The Infrared Space Observatory (ISO) mission, *Astronomy & Astrophysics*, **315**, L27-L31.
de Graauw, Th. et al. 1996, Observing with the ISO Short-Wavelength Spectrometer, *Astronomy & Astrophysics*, **315**, L49-L54.

Chapter 8: A Popular Account of Planetary Nebulae

Kwok, S. 2001, *Cosmic Butterflies*, Cambridge University Press

Chapter 8: A Technical Review of Planetary Nebulae

Kwok, S. 2000, *The Origin and Evolution of Planetary Nebulae*, Cambridge University Press

Chapter 8: Life of Lawrence Aller

DeVorkin, D. 1979, Oral History Transcript – Dr. Lawrence H. Aller, Niels Bohr Library & Archives, MD, USA

Chapter 8: *IRAS* Observations of Planetary Nebulae

Zhang, C.Y., Kwok, S. 1991, Spectral Energy Distribution of Compact Planetary Nebulae, *Astronomy & Astrophysics*, **250**, 179–211.

Chapter 8: Discovery and Observed Properties of Proto-Planetary Nebulae

Kwok, S. 1993: Proto-Planetary Nebulae, *Annual Review of Astronomy & Astrophysics*, **31**, 63–92.
Kwok, S. 2001: Proto-Planetary Nebulae, *Encyclopedia of Astronomy and Astrophysics*, Institute of Physics Publishing, p. 2168

Chapter 9: Detection of Carbon Monoxide in Carbon Stars

Solomon, P., Jefferts, K.B., Penzias, A.A., & Wilson, R.W. 1971, Observation of CO Emission at 2.6 millimeters from IRC+10216, *Astrophysical Journal*, **163**, L53-56.

Wilson, R.W., Solomon, P.M., Penzias, A.A., & Jefferts, K.B. 1971, Millimeter Observations of CO, CN, and CS Emission from IRC+10216, *Astrophysical Journal*, **169**, L35-L37.

Chapter 9: Laboratory Synthesis of Fullerene

Kroto, H.W., Heath, J.R., Obrien, S.C., Curl, R.F. & Smalley, R.E. 1985, C_{60}: Buckminsterfullerene. *Nature* **318**, 162–163.

Chapter 9: Astronomical Detection of Fullerene

Cami, J., Bernard-Salas, J., Peeters, E. & Malek, S.E. 2010, Detection of C_{60} and C_{70} in a Young Planetary Nebula. *Science*, **329**, 1180–1182.

García-Hernández, D. A. *et al.* 2010, Formation of Fullerenes in H-containing Planetary Nebulae. *Astrophysical Journal,***724**, L39-L43.

Sellgren, K. *et al.* 2010, C_{60} in Reflection Nebulae. *Astrophysical Journal,***722**, L54-L57.

Zhang, Y. & Kwok, S. 2010, Detection of C_{60} in tkhe Protoplanetary Nebula IRAS 01005+7910. *Astrophysical Journal,***730**, 126–130.

Chapter 9: Fullerene in Meteorites

Becker, L., Bunch, T.E. & Allamandola, L.J. 1999, Higher Fullerenes in the Allende Meteorite. *Nature,* **400**, 227–228.

Chapter 10: Chemical Properties of Soot

Harris, S.J., and Weiner, A.M. 1985, Chemical kinetics of Soot Particle Growth, *Annual Review of Physical Chemistry* **36**, 31–52.

Chapter 10: Discovery of UIR Features in a Planetary Nebula

Russell, R.W., Soifer, B.T., & Willner, S.P. 1977, The 4 to 8 μm Spectrum of NGC 7027, *Astrophysical Journal*, **217**, L149-L153.

Chapter 10: Identification of the Aromatic Nature of the UIR Features

Knacke, R.F. 1977, Carbonaceous Compounds in Interstellar Dust, *Nature*, **269**, 132–134.

Duley, W.W., & Williams, D.A., 1979, Are There Organic Grains in the Interstellar Medium? *Nature*, **277**, 40–41.
Duley, W. and Williams, D.A. 1981, The infrared Spectrum of Interstellar Dust: Surface Functional Groups on Carbon, *Monthly Notices of the Royal Astronomical Society*, **196**, 269–274.

Chapter 10: Infrared Properties of Polycyclic Aromatic Hydrocarbons

Allamandola, L.J., Gielens, G.G.M., Barker, J.R. 1989, Interstellar Polycyclic Aromatic Hydrocarbons - The infrared Emission Bands, the Excitation/Emission Mechanism, and the Astrophysical Implications, *Astrophysical Journal Supplements*, **71**, 733–775.
Cook, D.J., Schlemmer, S., Balucani, N., Wagner, D.R., Steiner, B., Saykally, R.J. 1996, Infrared Emission Spectra of Candidate Interstellar Aromatic Molecules. *Nature* **380**, 227–229.
Wagner, D.R., Kim, H. & Saykally, R.J. 2000, Peripherally Hydrogenated Neutral Polycyclic Aromatic Hydrocarbons as Carriers of the 3 Micron Interstellar Infrared Emission Complex: Results from Single-Photon Infrared Emission Spectroscopy. *Astrophysical Journal.* **545**, 854–860.

Chapter 10: Comments on the PAH Hypothesis

Donn, B.D., Allen, J.E., Khanna, R.K. 1989, A Critical Assessment of the PAH Hypothesis, in *IAU Symposium 135: Interstellar Dust*, eds. L.J. Allamandola and A.G.G.M. Tielens, Kluwer, p. 181–189.

Chapter 10: Detection of Aliphatic Compounds in Proto-Planetary Nebulae

Kwok, S., Volk, K., and Hrivnak, B.J. 1999: Chemical Evolution of Carbonaceous Materials in the Last Stages of Stellar Evolution, *Astronomy & Astrophysics*, **350**, L35-L38.
Kwok, S., Volk, K., and Bernath, P. 2001: On the Origin of Infrared Plateau Features in Proto-Planetary Nebulae, *Astrophysical Journal Letters*, **554**, L87-L90.

Chapter 10: Coal and Kerogen as Possible Carriers of the UIR Feature

Guillois, O., Nenner, I., Papoular, R., & Reynaud, C. 1996: Coal Models for the Infrared Emission Spectra of Proto-Planetary Nebulae, *Astrophysical Journal*, **464**, 810–817.
Papoular, R. 2001, The use of kerogen data in understanding the properties and evolution of interstellar carbonaceous dust. *Astronomy & Astrophysics*, **378**, 597–607.

Chapter 10: Properties of Terrestrial Kerogens

Vandenbroucke, M. 1980: Structure of Kerogens as seen by Investigations on Soluble Extracts, in *Kerogen*, Editions Technip, p. 415–444.

Chapter 10: Synthesis of Artificial Carbonaceous Compounds

Jones, A.P., Duley, W.W., and Williams, D.A. 1990, The Structure and Evolution of Hydrogenated Amorphous Carbon Grains and Mantles in the Interstellar Medium, *Quarterly Journal of the Royal Astronomical Society*, **31**, 567–582.

Sakata, A., Wada, S., Tanabè, T. & Onaka, T. 1984, Infrared Spectrum of the Laboratory-Synthesized Quenched Carbonaceous Composite (QCC): Comparison with the Infrared Unidentified Emission Bands, *Astrophysical Journal*, **287**, L51-L54.

Wada, S., Kaito, C., Kimura, S., Ono, H. & Tokunaga, A. T. 1999, Carbonaceous Onion-like Particles as a Component of Interstellar Dust. *Astronomy & Astrophysics*, **345**, 259–264

Wada, S., Tokunaga, A. 2006, Carbonaceous Onion-like Particles: a Possible Component of the Interstellar Medium, in *Natural Fullerences and Related Structures of Elemental Carbon*, F.J. M. Rietmeijer (ed.), Springer, p. 31–52.

Rotundi A, Rietmeijer F.J.M., Colangeli L, Mennella V, Palumbo P, & Bussoletti E 1998: Identification of Carbon Forms in Soot Materials of Astrophysical Interest. *Astronomy & Astrophysics*, **329**, 1087–1096

Pino T, Dartois E, Cao A-T, Carpentier Y, Chamaillé T, Vasquez R, Jones AP, D'Hendecourt L, & Bréchignac P 2008: The 6.2 μm Band Position in Laboratory and Astrophysical Spectra: a Tracer of the Aliphatic to Aromatic Evolution of Interstellar Carbonaceous Dust. *Astronomy & Astrophysics*, **490**, 665–672

Chapter 10: The Life of Akira Sakata

Tokunaga, A.T. 1987, Akira Sakata and Quenched Carbonaceous Composite, in *From Stardust to Planetesimals*, ASP conference series, Vol. 122, eds. Y.J. Pendleton and A.G.G.M. Tielens, p. 17.

Chapter 10: Mixed Aromatic/Aliphatic Nanoparticles as Carriers of UIR Features

Kwok, S., Zhang, Y. 2011, Mixed Aromatic-Aliphatic Organic Nanoparticles as Carriers of Unidentified Infrared Emission Features, *Nature*, **479**, 80–83.

Chapter 11: Condensation of Minerals in the Atmospheres of Stars

Gilman, R.C. 1974, Planck Mean Cross-Sections for Four Grain Materials, *Astrophysical Journal Supplements.*, **28**, 397–403.

Chapter 11: Laboratory Studies of Corundum

Begemann B, et al. 1997, Aluminum Oxide and the Opacity of Oxygen-rich Circumstellar Dust in the 12–17 Micron Range, *Astrophysical Journal*, **476**, 199–208.

Chapter 11: The 13-μm Unidentified Feature

Little-Marenin, I.R., Little, S.J. 1988, Emission features in IRAS low-resolution spectra of MS, S and SC stars, *Astrophysical Journal*, **333**, 305–315.
Onaka, T., de Jong, T., and Willems, F.J. 1989, A Study of M-Mira Variables Based on IRAS LRS Observations .1. Dust Formation In The Circumstellar Shell.,*Astronomy & Astrophysics*, **218**, 169–179.
Posch T et al. 1999, On the origin of the 13 μm feature. A study of ISO-SWS spectra of oxygen-rich AGB stars, *Astronomy & Astrophysics*, **352**, 609–618.

Chapter 11: GEMS

Bradley, J.P. et al. 1999, An Infrared Spectral Match between GEMS and Interstellar Grains. *Science*, **285**, 1716–1718.

Chapter 11: A Summary of the New Field of Astromineralogy

Henning, Th. (editor), 2003, *Astromineralogy*, Springer-Verlag (Heidelberg)

Chapter 12: Laboratory Synthesis of Nanodiamonds

Kimura, Y., Sato, T., Kaito, C. 2005, Production of Diamond and Solid-solution Nanoparticles in the Carbon-Silicon System using Radio-frequency Plasma, *Carbon*, **43**, 1557

Chapter 12: Nanodiamonds as Evidence for Past Major Impact Events

Israde-Alcántara, I. *et al.* 2012, Evidence from Central Mexico Supporting the Younger Dryas Extraterrestrial Impact Hypothesis. *Proceedings of the National Academy of Sciences* 109: E738-E747
Fiedel, S.J. 2011, The Mysterious Onset of the Younger Dryas, *Quaternary International* **242**, 262–266.

Chapter 12: Detection of Interstellar Nanodiamonds in Meteorites

Lewis, R.S., Tang, M., Wacker, J.F., Anders, E., Steel, E. 1987, Interstellar Diamonds in Meteorites, *Nature*, **326**, 160–162.

Chapter 12: Discovery of the 21-μm Feature

Kwok, S., Volk, K.M. & Hrivnak, B.J. 1989, A 21 Micron Emission Feature in Four Proto-Planetary Nebulae. *Astrophysical Journal*, **345**, L51-L54.

Chapter 12: Suggestion of Diamond as the Carrier of the 21-μm Emission Feature

Hill, H.G.M., Jones, A.P., and d'Hendecourt, L.B. 1998, Diamonds in Carbon-rich Proto-Planetary Nebulae, *Astronomy & Astrophysics*, **336**, L41-L44.

Chapter 12: Suggestion of Hydrogenated Fullerene as the Carrier of the 21-μm Emission Feature

Webster, A. 1995, The Lowest of the Strongly Infrared Active Vibrations of the Fulleranes and Astronomical Emission band at a Wavelength of 21-microns, *Monthly Notices of the Royal Astronomical Society*, **277**, 1555–1566.

Chapter 12: Suggestion of Hydrogenated TiC as the Carrier of the 21-μm Emission Feature

von Helden G., Tielens, A.G.G.M., van Heijnsbergen, D., Duncan, M.A., Hony, S., Waters, L.B.F. M., Meijer, G. 2000, Titanium Carbide Nanocrystals in Circumstellar Environments, *Science*, **288**,313–316

Chapter 12: Classification of the UIR Features

Geballe, T.R. 1997, Spectroscopy of the Unidentified Infrared Emission Bands, in *From Stardust to Planetesimals*, ASP Conference Series, Vol. **122**, ed. Yvonne J. Pendleton; A.G.G.M. Tielens, p.119–128, Astronomical Society of Pacific, San Francisco, CA

Chapter 12: Laboratory Studies of Diamond

Chen, C.F., Wu, C.C., Cheng, C.L., Sheu, S.Y. & Chang, H.C. 2002, The Size of Interstellar Nanodiamonds Revealed by Infrared Spectra of CH on Synthetic Diamond Nanocrystal Surfaces. *Journal of Chemical Physics*, **116**, 1211–1214 and references therein

Chapter 12: Identification of Interstellar Diamond

Guillois, O., Ledoux, G., Reynaud, C. 1999, Diamond Infrared Emission Bands in Circumstellar Media, *Astrophysical Journal*, **521**, L133-136.
Van Kerckhoven, C., Tielens, A. & Waelkens, C. 2002, Nanodiamonds around HD 97048 and Elias 1. *Astronomy & Astrophysics*, **384**, 568–584.

Chapter 13: Discovery of the Red Rectangle

Cohen, M et al. 1975, The Peculiar Object HD 44179 ("The Red Rectangle"), *Astrophysical Journal,* **196**, 179–189.
Schmidt, G.D., Cohen, M., & Margon, B. 1980, Discovery of Optical Molecular emission from the Bipolar Nebula surrounding HD 44179, *Astrophysical Journal,* **239**, L133-L138.

Chapter 13: Bioluminescence

Harvey, E.N. 1952, *Bioluminescence,*. Academic Press, New York
Lee, J. 2008, Bioluminescence: the first 3000 years, *Journal of Siberian Federal University, Biology,* **3**, 194–205.

Chapter 13: Diamond as the Carrier of the ERE

Chang, H.-C., Chen, K., and Kwok, S. 2006, Nanodiamonds as a Possible Carrier of Extended Red Emission, *Astrophysical Journal Letters,* **639**, L63-L66.

Chapter 14: On the Origin of Oil

Robinson, R. 1966, The Origins of Petroleum. *Nature* **212**, 1291–1295
Walters, C. 2006, The Origin of Petroleum, *Practical Advances in Petroleum Processing,* ed. C. Hsu & P. Robinson, 79–101.

Chapter 14: The Link of Petroleum to Life

Muniyappan, R. 1955, Porphyrins in petroleum, Journal of Chemical Education, 32 (5), 277–279.

Chapter 14: The Abiogenic Theory of Oil

Gold, T. 1987: *Power from the Earth,* J.M. Dent & Sons
Gold, T. 1999: *The Deep Hot Biosphere,* Springer-Verlag (in paperback, 2001, Copernicus Books).
Helgeson, H.C., Richard, L., McKenzie, W.F., Norton, D.L., Schmitt, A. 2009, A Chemical and Thermodynamic Model of Oil Generation in Hydrocarbon Source Rocks, *Geochimica et Cosmochimica Acta,***73**, 594–695.
Kenney, J.F., Kutcherov, V.A., Bendeliani, N.A., Alekseev, V.A. 2002, The Evolution of Multi-component Systems at High Pressures: VI. The Thermodynamic Stability of the Hydrogen Carbon System: the Genesis of Hydrocarbons and the Origin of Petroleum, *Proceedings of the National Academy of Sciences of the USA,* **99**, 10976–10981.

Sherwood Lollar, B., Westgate, T.D., Ward, J.A., Slater, G.F., Lacrampe-Couloume, G. 2002, Abiogenic Formation of Alkanes in the Earth's Crust as a Minor Source for Global Hydrocarbon Reservoirs, *Nature*, **416**, 522–524.

Chapter 14: Life of Tommy Gold

Gold, T. (Simon Mitton, editor) 2012, *Taking the Back off the Watch*, Springer.

Chapter 14: Organics in Titan

Lorenz, R.D., Mitchell, K.L., Kirk, R.L., Hayes, A.G., Aharonson, O., Zebker, H.A., Paillou, P., Radebaugh, J., Lunine, J.I., Janssen, M.A., Wall, S.D., Lopes, R.M., Stiles, B., Ostro, S., Mitri, G., Stofan, E.R. 2008, Titan's Inventory of Organic Surface Materials, *Geophysical Research Letters*, **35**, L02206.

Chapter 14: Petroleum in Space?

Cataldo F, Keheyan Y 2003, Heavy Petroleum Fractions as Possible Analogues of Carriers of the Unidentified Infrared Bands, *International Journal of Astrobiology*, **2**, 41–50
Cataldo F, Keheyan Y, Heymann D 2004, Complex Organic Matter in Space: About the Chemical Composition of Carriers of the Unidentified Infrared Bands (UIBs) and Protoplanetary Emission Spectra Recorded from Certain Astrophysical Objects. *Origins of Life and Evolution of the Biosphere*, **34**, 13–24.

Chapter 14: Inventory of Carbon on Earth

Falkowski, P., Scholes, R.J., Boyle, E., Canadell, J., Canfield, D., Elser, J., Gruber, N., Hibbard, K., Högberg, P., Linder, S., Mackenzie, F.T., Moore, B., Pedersen, T., Rosenthal, Y., Seitzinger, S., Smetacek, V., & Steffen, W. 2000, The Global Carbon Cycle: A Test of our Knowledge of Earth as a System, *Science*, **290**, 291–296.
Lunine, J.I. 2005, *Astrobiology*, Addison Wesley, Boston

Chapter 14: The Giant Impact Theory of the Origin of the Moon

Hartmann, W.K., and Davis, D.R. 1975, Satellite-sized Planetesimals and Lunar Origin, *Icarus*, **24**, 504–514.
Hartmann, W.K., Phillips, R.J., Taylor, G.J. (editors), 1986, *Origin of the Moon*, Lunar and Planetary Institute, Houston

Chapter 14: Delivery of Stellar Hydrocarbons to the Solar System

Kwok, S. 2009, Delivery of Complex Organic Compounds from Planetary Nebulae to the Solar System, *International Journal of Astrobiology*, **8**, 161–167.

Chapter 14: Search for Oil on Mars

Maurette, M., Brack, A., Duprat, J., and Engrand, C. 2006, Kerogen-rich Micrometeorites and
Crude Petroleum in Hadean Time, Lunar and Planetary Science XXXVII, abstract no. 1583
Direito, M.S., and Webb, M.E. 2006, Search for Oil Reserves on Mars, paper presented at the 6th
European Workshop on Astrobiology, Lyon
McGowan, J.F. 2000, Oil and Natural Gas on Mars, *Instruments, Methods, and Missions for
Astrobiology III, Proceedings of Photo-Optical Instrumentation Engineers*, **4137**, 63–74.

Chapter 15: Infrared Detection of Organic Molecules in Planets

S.T. Ridgway, H.P. Larson, & U. Fink. 1976, The infrared spectrum of Jupiter. In Proceedings of
the Colloquium, Tucson, Arizona, May 19–21, 1975, *Jupiter: Studies of the interior, atmosphere, magnetosphere, and satellites*

Chapter 15: Laboratory Synthesis of Tholins

Sagan, C., & Khare, B.N. 1979, Tholins: Organic Chemistry of Interstellar Grains and Gas,
Nature, **277**, 102–107.

Chapter 15: Organics in Asteroids

Gradie, J., & Veverka, J. 1980, The composition of the Trojan asteroids, *Nature*, **283**, 840–842.

Chapter 15: Discovery of Organics in the Orgueil Meteorite

Hodgson, G. W. & Baker, B. L. 1964, Evidence for Porphyrins in the Orgueil Meteorite. *Nature*
202, 125–131.
Nagy, B., Murphy, M.T.J. & Modzeleski, V.E. 1964, Optical Activity in Saponified Organic
Matter Isolated from the Interior of the Orgueil Meteorite. *Nature* **202**, 228–233.
Baker, B.L. 1971, Review of Organic Matter in the Orgueil Meteorite. *Space Life Sciences* **2**,
472–497.

Chapter 15: Organics in the Murchison and Other Meteorites

Cronin, J.R., Pizzarello, S., Cruikshank, D.P. 1988, Organic Matter in Carbonaceous Chondrites,
Planetary Satellites, Asteroids and Comets, in *Meteorites and the Early Solar System* (eds. John
F. Kerridge & M.S. Mathews), 819–857.
Schmitt-Kopplin, P. *et al.* 2010, High Molecular Diversity of Extraterrestrial Organic Matter in
Murchison Meteorite Revealed 40 Years after its Fall. *Proceedings of the National Academy of
Sciences of the USA*, **107**, 2763–2768.
Nakamura-Messenger, K., Messenger, S., Keller, L.P., Clemett, S.J., Zolensky, M.E. 2006,
Organic globules in the Tagish Lake Meteorite: remnants of the protosolar disk, *Science*,
314, 1439–1442.
Martins, Z., et al. 2008, Extraterrestrial Nucleobases in the Muchison Meteorite, *Earth & Plane-
tary Science Letters*, **270**, 130–136.

Ehrenfreund, P., Robert, F., d'Hendecourt, L., Behar, F. 1991, Comparison of Interstellar and Meteoritic Organic Matter at 3.4 micron, *Astronomy & Astrophysics*, **252**, 712–717.

Martins, Z., O'D Alexander, C.M., Orzechowska, G.E., Fogel, M.L., Ehrenfreund, P. 2007, Indigenous amino acids in primitive CR meteorites, *Meteoritics & Planetary Science* **42**, Nr 12, 2125–2136

Cody, G.D., Heying, E., Alexander, C.M.O., Nittler, L.R., Kilcoyne, A.L.D., Sandford, S.A., Stroud, & R.M. 2001, Cosmochemistry Special Feature: Establishing a Molecular Relationship between Chondritic and Cometary Organic Solids. *Publication National Academy of Sciences of the USA 108(9):3516-21*

Callahan, M. P. *et al.* 2011, Carbonaceous Meteorites Contain a Wide Range of Extraterrestrial Nucleobases. *Proceedings of the National Academy of Sciences of the USA*, **108**, 13995–13998.

Pizzarello, S., Williams, L. B., Lehman, J., Holland, G. P., & Yarger, J. L. 2011, *Proceedings of the National Academy of Sciences of the USA*, **108**, 4303–4306.

Chapter 15: Organics in Interplanetary Dust Particles

Brownlee, D. E. 1978, in *Cosmic Dust*, ed. J. A. M. McDonnell, (N.Y.: J.Wiley), 295

Kerridge J. F. 1999 Formation and Processing of Organics in the Early Solar System, *Space Science Reviews* **90**, 275–288.

Flynn, G. J.; Keller, L. P.; Feser, M.; Wirick, S.; Jacobsen, C. 2003 The Origin of Organic Matter in the Solar System: Evidence from the Interplanetary Dust Particles *Geochimica et Cosmochimica Acta*, **67**, 4791–4806.

Chapter 15: Plumes from Enceladus

Waite, J. H. *et al.* 2006, Cassini Ion and Neutral Mass Spectrometer: Enceladus Plume Composition and Structure, *Science*, **311**, 1419–1422.

Chapter 16: Diamond in Meteorites

Lewis, R.S., Tang, M., Wacker, J.F., Anders, E., & Steel, E. 1987, Interstellar Diamonds in Meteorites, *Nature*, **326**, 160–162.

Chapter 16: Discovery of Presolar Grains

Bernatowicz, T. *et al.* 1987, Evidence for Interstellar SiC in the Murray Carbonaceous Meteorite. *Nature* **330**, 728–730.

Zinner, E. 1998, Stellar Nucleosynthesis and the Isotopic Composition of Presolar Grains from Primitive Meteorites. *Annual Review of Earth and Planetary Science*, **26**, 147–188

Bernatowicz, T.J., Zinner, E. 1997, *Astrophysical Implications of the Laboratory Study of Presolar Materials*, American Institute of Physics (New York)

Nittler, L.R., Alexander, C.M., Gao, X., Walker, R.M., Zinner, E. 1997, Stellar Sapphires: The Properties and Origins of Presolar Al_2O_3 in Meteorites, *Astrophysical Journal*, **483**, 475–495.

Messenger, S., Keller, L. P., Stadermann, F. J., Walker, R. M. & Zinner, E. 2003, Samples of Stars Beyond the Solar System: Silicate Grains in Interplanetary Dust, *Science*, **300**, 105–108.

Davis, A.M. 2011, Stardust in Meteorites, *Proceedings of the National Academy of Sciences of the USA*, **108**, 19142–19146.

Chapter 17: Search for Prebiotic Molecules in Space

Hollis, J.M., Lovas, F.J., & Jewell, P.R. 2000, Interstellar Glycolaldehyde: the First Sugar, *Astrophysical Journal*, **540**, L107-110.

Kuan, Y.-J., Charnley, S. B., Huang, H.-C., Tseng, W.-L., & Kisiel, Z. 2003, Interstellar Glycine, *Astrophysical Journal*, **593**, 848–867.

Kuan, Y.J., Yan, C.-H., Charnley, S.B., Kisiel, Z., Ehrenfreund, P., & Huang, H.-C. 2003, A search for interstellar pyrimidine, *Monthly Notices of the Royal Astronomical Society*, **345**, 650–656.

Chapter 17: Organics as a Component of Interstellar Grains

Hoyle, F. & Wickramasinghe, N. C. 1977, Polysaccharides and Infrared Spectra of Galactic Sources, *Nature*, **268**, 610–612

Duley, W. W. & Williams, D. A. 1979, Are There Organic Grains in the Interstellar Medium? *Nature*, **277**, 40–41.

Chapter 17: Discovery of Organics in the Diffuse Interstellar Medium

Wickramasinghe, D. T. & Allen, D. A. 1980, The 3.4-micron Interstellar Absorption Feature. *Nature*, **287**, 518–519.

Chapter 17: Bacteria and Archaea as Domains of Life

Roussel, E.G., Bonavita, M.-A.C., Querellou, J., Cragg, B.A., Webster, G., Prieur, D., & Parkes, R. J. 2008, Extending the Sub-Sea-Floor Biosphere, *Science* **320**, 1046.

Woese, C.R., and Fox, G.E. 1977, Phylogenetic Structure of the Prokaryotic Domain: the Primary Kingdom, *Proceedings of the National Academy of Sciences of the USA*, **74**, 5088–5090.

Chapter 17: The Deep Biosphere

Røy, H. *et al.* 2012, Aerobic Microbial Respiration in 86-Million-Year-Old Deep-Sea Red Clay, *Science* **336**, 922–925.

Chapter 17: Phosphorus

Maciá, E. 2005, The Role of Phosphorus in Chemical Evolution, *Chemical Society Reviews*, **34**, 691–701

Chapter 17: Interstellar Transport of Bacteria

Valtonen, M., Nurmi, P., Zheng, J.-Q, Cucinotta, F.A., Wilson, J.W., Horneck, G, Lindegren, L, Melosh, J., Rickman, H., Mileikowsky, C. 2009, Natural Transfer of Viable Microbes in Space from Planets in Extra-Solar Systems to a Planet in our Solar System and Vice Versa. *Astrophysical Journal*, **690**, 210–215

Chapter 18: A General Overview of Comets

Fernández, J.A. 2005, *Comets: Nature, Dynamics, Origin, and their Cosmognoical Relevance*, Springer.

Chapter 18: Ancient Observations of the Halley's Comet

Tsu, W.S. (1934) The observations of Halley's comet in Chinese history, *Popular Astronomy*, **42**, 191–201

Chapter 18: Comets and Life on Earth

Hoyle, F., Wickramasinghe, N.C. 1999, Comets – a Vehicle for Panspermia, *Astrophysics & Space Science*, **268**, 333–341.
Oró, J. 1961, Comets and the Formation of Biochemical Compounds on the Primitive Earth, *Nature*, **190**, 389–90.
Delsemme, A. 1998, *Our Cosmic Origins: from the Big Bang to the emergence of life and intelligence*, Cambridge University Press

Chapter 18: Organics in Comets

Sandford, S. et al. 2006, Organics Captured from Comet 81P/Wild 2 by the Stardust Spacecraft, *Science*, **314**, 1720–1724.
Mumma, M.J., et al. 2001, A Survey of Organic Volatile Species in Comet C/1999 H1 (Lee) using NIRSPEC at the Keck Observatory, *Astrophysical Journal*, **546**, 1183–1193.
Mumma, M.J., DiSanti, M.A., Dello Russo, N., Magee-Sauer, K., Gibb, E., & Novak, R. 2003, Remote Infrared Observations of Parent Volatiles in Comets: a Window on the Early Solar System, *Advances in Space Research*, **31**, 2563–2575

Chapter 19: Water as an Element of Life

Kasting, J. F. and Catling, D. 2003, Evolution of a Habitable Planet. *Annual Review of Astronomy & Astrophysics* **41**, 429–463.

Chapter 19: The Origin of Water on Earth

Mottl, M., Glazer, B., Kaiser, R. & Meech, K. 2007, Water and Astrobiology. *Chemie der Erde / Geochemistry* **67**, 253–282.

Chapter 19: Comets as Sources of Terrestrial Water

Chyba, C.F. 1990, Impact Delivery and Erosion of Planetary Oceans in the Early Inner Solar System, *Nature*, **343**, 129–133
Delsemme, A. 1996 *The origin of the atmosphere and of the oceans*, Comets and the Origin and Evolution of Life, eds. P.J. Thomas, C.F. Chyba, C.P. McKay, Springer, p. 29

Chapter 19: Discovery of Main Belt Comets

Hsieh, H.H., and Jewitt, D.C. 2006, A Population of Comets in the Main Asteroid Belt, *Science*, **312**, 561–563

Chapter 19: Effects of Cometary Impacts

Napier, W.M., and Clube, S.V.M. 1979, A Theory of Terrestrial Catastrophism, *Nature*, **282**, 455–459
Oró, J. 1961 Comets and the Formation of Biochemical Compounds on the Primitive Earth, *Nature*, **190**, 389–390

Chapter 19: The GAIA Hypothesis

Lovelock, J. 1979, *GAIA: A New Look at the Life on Earth*, Oxford University Press, Oxford
Lovelock, J. 1988 *The Ages of GAIA: a Biography of our Living Earth*, Oxford University Press

Chapter 20: Miller-Urey Type Experiments

Parker, E. T. *et al.* 2011, Primordial Synthesis of Amines and Amino acids in a 1958 Miller H_2S-rich Spark Discharge Experiment. *Proceedings of the National Academy of Sciences of the USA*, 108(14), 5526-31 doi:10.1073/pnas.1019191108.

Chapter 20: Life of Mayo Greenberg

Allamanandola, L.J., d'Hendecourt, L. 2004, A tribute to the Life and Science of J. Mayo Greenberg, in *Astrophysics of Dust*, ASP conf. ser. Vol. 309, eds. A.N. Witt, G.C. Clayton, B.T. Draine, p. 1

Chapter 20: Simulation of the Synthesis of Organics Under Interstellar Conditions:

Bernstein, M.P., Dworkin, J.P., Sandford, S.A., Cooper, G.W., & Allamandola, L.J. 2002, Racemic Amino Acids from the Ultraviolet Photolysis of Interstellar Ice Analogues, *Nature*, **416**, 401–403.
Muñoz Caro, G.M., et al., 2002, Amino Acids from Ultraviolet Irradiation of Interstellar Ice Analogues, *Nature*, **416**, 403–406.
Kobayashi, K. , Takano, Y., Masuda, H., Tonishi, H., Kaneko, T., Hashimoto, H. & Saito, T. 2004, Possible cometary organic compounds as sources of planetary biospheres, *Advances in Space Research*, **33**, 1277–1281

Chapter 21: UIR Features in the Diffuse Interstellar Space

Tanaka, M. *et al.* 1996, IRTS Observation of the Unidentified 3.3-Micron Band in the Diffuse Galactic Emission. *Publications of the. Astronomical Society Japan,* **48**, L53-L57.

Onaka, T., Yamamura, I., Tanabe, T., Roellig, T. L. & Yuen, L. M. Detection of the mid-infrared unidentified bands in the diffuse galactic emission by IRTS. *Publications of the. Astronomical Society Japan.* **48**, L59-L63.

Chapter 21: UIR Features in Reflection Nebulae

Uchida, K. I., Sellgren, K., Werner, M. W. & Houdashelt, M. L. 2000, Infrared Space Observatory mid-infrared spectra of reflection nebulae. *Astrophysical Journal,* **530**, 817–833.

Chapter 21: UIR Features in Galaxies

Smith, J. D. T. *et al.* 2007. The mid-infrared spectrum of star-forming galaxies: Global properties of polycyclic aromatic hydrocarbon emission. *Astrophysical Journal,* **656**, 770–791.

Chapter 21: UIR Features in NGC 300

Prieto, J. L., Sellgren, K., Thompson, T. A. & Kochanek, C. S. 2009, A Spitzer/IRS Spectrum of the 2008 Luminous Transient in NGC 300: Connection to Proto-Planetary Nebulae. *Astrophysical Journal,* **705**, 1425–1432

Chapter 21: Stellar Winds from Old Stars

Kwok, S. 1975, Radiation Pressure on Grains as a Mechanism for Mass Loss in Red Giants, *Astrophysical Journal,* 198, 583–591.

Kwok, S. 1987, Effects of Mass Loss on the Late Stages of Stellar Evolution, *Physics Reports,* **156**, No. 3, p. 111–146.

Chapter 21: Organic Enrichment of the Solar System by Stardust

Kwok, S. 2004, The Synthesis of Organic and Inorganic Compounds in Evolved Stars, *Nature,* **430**, 985–991.

Ehrenfreund, P. et al. 2004, *Astrobiology: future perspectives*, Kluwer, Dordrecht.

Ehrenfreund, P. et al. 2002, Astrophysical and Astrochemical Insights into the Origin of Life, *Reports of Progress in Physics,* **65**, 1427–1487

Chapter 22: Age of the Earth

Burchfield, J. D. 1998. The age of the Earth and the Invention of Geological Time, in *Lyell: the Past is the Key to the Present*, eds. D.J. Blundell & A.C. Scott, Geological Society. London, Special Publications, 143, 137–143.

Index

21-μm emission feature, 116, 117
3.4-μm feature, 159–161, 183
11.3-μm SiC feature, 60, 68
10-μm silicate feature, 57, 64, 67
220 nm feature, 92, 102

A

AAT. See Anglo Australia Telescope (AAT)
Absolute zero, 47
Acetaldehyde, 154, 172
Acetamide, 185
Acetylene, 144, 154, 172, 183
Acidophiles, 157
Adenine, 139–141, 155
Aerobee rocket, 92
AFGL 3068, 50, 68
AFGL rocket survey, 68
AGB. See Asymptotic Giant Branch (AGB), 192
Alais meteorite, 8
Alanine, 4, 185
Alchemy, 121
Alcohols, 139, 172, 183
Algae, 128
Aliphatic compounds, 18, 97–99, 102, 117, 131, 133, 139, 141, 142, 151, 156, 159–161, 183
Allamandola, L., 95, 185
Allen, D., 159
Allende meteorite, 17, 87, 151
Altair, 49
Aluminum oxides, 108
Alvarez, W., 32
Amber, 7, 107
Amino acids, 5, 6, 19, 95, 139, 143, 151, 183, 185, 186, 195
Ammonia, 4–6, 82, 137, 156, 171, 173, 179, 183, 185

Amorphous materials, 101
Anaxagoras, 6
Anders, E., 147
Andromeda Galaxy, 217
Anglo Australia Telescope (AAT), 159
Animal fats, 98
Antarctica, 15, 16, 151
Anthracene, 96
Apollo mission, 11, 28, 205
Aquatic biosphere, 132
Archaea, 158, 161, 195
Aristotle, 15, 121, 165, 166, 203, 206
Aromatic compounds, 18, 95, 97, 115, 117, 156, 189, 192
Arrhenius, S., 7, 8
Asteroid belt, 11, 179
Asteroid 25143 Itokawa, 12
Asteroid 2685 Masursky, 142
Asteroids, xvi, 11, 13, 16, 21, 22, 27–29, 53, 54, 114, 138, 141, 145, 169, 170, 174, 179, 189, 194, 198, 201, 223
Astrobiology, 1
Astrochemistry, 1, 86
Astrology, 37, 41
Astromineralogy, 60, 111
Asymptotic Giant Branch (AGB), 63, 78, 189, 192, 194
Atacama large millimeter/submillimeter array (ALMA), 208
Atacama Pathfinder Experiment (APEX), 85
Atacama Submillimeter Telescope Experiment (ASTE), 85
Atmosphere, 5, 6, 11, 13–16, 21, 31, 32, 49, 56, 57, 65, 66, 73, 81, 85, 88, 92, 94, 113, 114, 117, 131, 133, 150, 152, 156, 158, 165, 174, 175, 177, 179, 180, 183